The New Scientific Angling

Trout and Ultraviolet Vision

Explored in depth by

Reed F. Curry

This work has come to fruition through the kind assistance of my patient wife, Jayne, who saw every room in our home temporarily transformed into a laboratory … and encouraged me nonetheless.

Some of the material from this book was built upon work written by the author for his website www.overmywaders.com and his blog 'The Contemplative Angler.'

Published by
Buckram Publishing
P.O. Box 493, Wilton, NH 03086
www.BuckramPublishing.com
ISBN 9780984086306

Cover design and layout by Anna Lightfoot.

Printed in the United States of America.

TABLE OF CONTENTS

INTRODUCTION

The evening hatch of small duns is coming off the water, a slow start but progressing well. In the last of the light you struggle to tie on a fly, your hands shaking slightly. You have to smile; despite all the years of fishing, the evening hatch still gives you buck-fever. Ten yards down-pool an old friend is working out line for his first cast. The sun is lost in the pines behind you. A light wind moves upstream, just enough to keep the mosquitoes at bay. In a few minutes two lines are arching over the pool, the familiar rhythm placing the flies in the paths of cruising trout.

An hour later you're both back at the car, breaking down your rods. Your friend took some beautiful fish, you hadn't a rise. Yet you had both fished the same size, pattern, and color fly. The only difference—there was no visible difference—was in the dubbing used in the body; yours was gray muskrat fur, his was rabbit fur dyed gray.

Invisible to you, your friend's fly reflected the ultraviolet markings the trout expected. And that was all the difference the trout required.

The literature of fly fishing for trout is rich with insights into the nature of the fish, its habits, and its capabilities. Progressive discoveries in the sciences—biology, optics, fluid dynamics—have provided anglers with tremendous detail into the feeding habits of trout. This in turn created opportunity for what has become known as "scientific angling," or strict imitation of the insects forming the trout's diet based upon known science.

From Art Flick's groundbreaking work *A New Streamside Guide to Naturals and their Imitations* of the 1960s through Clarke and Goddard's excellent *The Trout and the Fly* (1995), with many other valuable tomes before and since, the premise of imitating the natural fly has always been that trout will see the fly as we do.

That premise is false.

Trout do not see the world as we do, but, in a sense, far better. Within the past few years, scientists have discovered that trout, like many other creatures, can utilize the ultraviolet range of light—specifically, the long wavelength ultraviolet—for detecting food, prey, danger, and mates. Close examination revealed that the trout eye contains, in addition to the three color cones of humans—red, green, and blue—an ultraviolet-sensitive cone.[1][2][3]

Anglers have always been an insightful lot. We always knew that trout have extraordinary vision. Sixty years ago, one fishing writer recorded his suspicion that trout could see in the ultraviolet:

> "... experiments have been carried right through the spectrum and indicate that fish in general can not only distinguish pure colors from one another, but also are adept at spotting the various shades. Moreover, their eyes are *apparently more susceptible to colors at the ends of the spectrum—the ultraviolets and infrareds—than are human eyes.* Experiments have also indicated that color sensitivity tends to diminish with fading light, which is hardly surprising as our eyes register the same diminution.
>
> ...Now, I don't know anything about the night vision of trout, except that it is uncanny by our own standards, but an analogy might be drawn between a trout's eyes and a snooperscope. *If a trout's vision extends beyond ours into the ends of the spectrum, this may account for his ability to see at night.* It is a known fact that certain objects—the wings of moths, the fur of some animals, various insects—fluoresce when exposed to ultraviolet light. Possibly a trout can see some such things which we can't. I don't know. [emphasis mine]

From *Trout Fishing* by Dan Holland (Crowell Co., NY, 1949)

In 1994, John Merwin produced the superb work *The New American Trout Fishing* (Macmillan Publishing, NY, 1994) wherein he writes:

> "We are left with a very large question. While we know that colors can be seen, we still have little idea as to how they are perceived by trout. I suspect the ringer in this whole business is the trout's demonstrated ability to see the ultraviolet portion of the spectrum in addition to those longer wavelengths and colors of light that are visible to humans. The trout's ultraviolet sensitivity has been known for some time in the scientific community and certainly isn't radical news, but it's never been adequately related to our many theories of artificial-fly dressing...
>
> ...We both see a color, but apparently we and the trout do not see the same color at the same time.
>
> Thinking about this has given me a headache more than once, for reasons readily apparent to any trout fisherman of some experience. The whole idea throws many imitative concepts into a cocked hat because we can quite literally no longer see what we're doing, at least from the trout's viewpoint."

As noted by one fisheries biologist who is also a fisherman:

> If trout have the ability to see further into the UV an [sic] IR spectrums than humans can, is it possible that there are colors and materials unnoticeable to us that trout might prefer? It is known that many feathers reflect different colors of light unseen to the human eye. Have you ever wondered why flies that appear to be the same have different affects [sic] on feeding trout?
>
> Perhaps they are tied with similar but not identical materials, emitting many different colors underwater—Food for Thought—[4]

This book introduces the results of recent scientific research in the field of ultraviolet vision and casts that in the light of its impact on fly fishing for trout. Many of the following pages represent original research, allowing the reader to "see" aquatic insects, baitfish, fly-tying materials, and artificial flies through the medium of ultraviolet reflectance photography. The discerning angler will find in this a new approach to designing and fishing flies for trout.

I have endeavored throughout this work to provide references to peer-reviewed scientific studies. In order to keep the text readable this supporting evidence has been placed in end-notes.

I should warn the reader that I have allowed my altered-ego Overmywaders to interject his own thoughts—if that term is applicable—and theories in this work. However, these have been carefully relegated to the margins.

CHAPTER ONE

A LIGHT INTRO TO LIGHT

In order to appreciate ultraviolet vision, we should initially understand some of the properties of light. What we perceive as light is just a small section of the electromagnetic spectrum, which ranges from microwaves to gamma waves. The wavelength of electromagnetic radiation is generally measured in nanometers (nm), billionths of a meter. The wavelength of the light most humans can see is in the range of, approximately, 400–700nm. This is known as the "visible spectrum." Below those wavelengths we have the Ultraviolet spectrum which is subdivided into categories according to the shorter and shorter wavelengths. The invisible spectrum of concern to us is that from 320-400nm, known as UVA. It is within this range—optimally at 365nm—that trout discern features of their environment that are invisible to us.

Since trout have vision capabilities in the UVA, one question should be "How much UVA is available to trout?" The short-wave radiation from the sun striking the Earth is, on average, 8% UV, 39% visible, and 53% infrared.[5] Most of that UV is UVA. However, on a cloudy day the percentage of UV changes. UVA is less absorbed by clouds than visible light, so the percentage of UV increases under these conditions. And at dusk the total amount of light

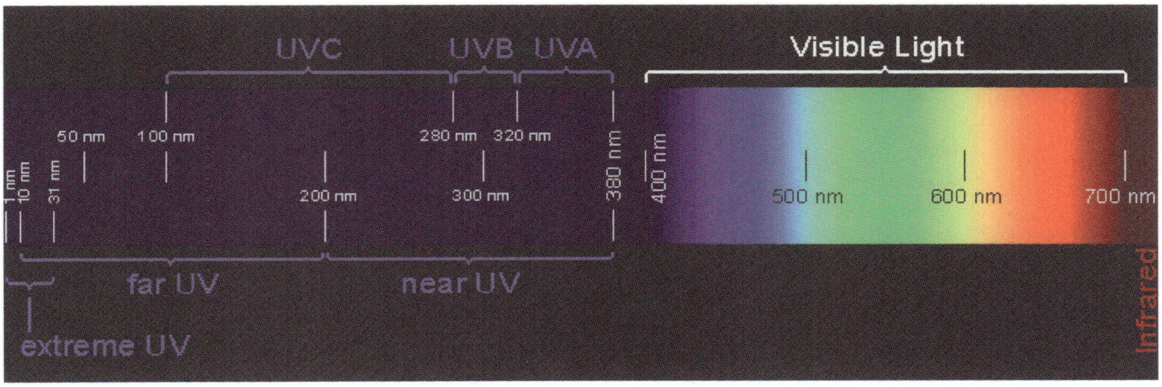

decreases, but again, the percentage of visible light decreases more than the percentage of UV light. This would indicate that on cloudy days and at dawn and dusk, trout may be more reliant on their UV vision.

Reflectance vs Fluorescence

There is widespread confusion regarding the terms "UV reflectance" and "UV fluorescence." UV fluorescence occurs when ultraviolet light strikes a material and triggers an emission of light at a longer wavelength—usually a visible wavelength. If you have seen a white shirt glow under a blacklight (UV light), you are observing UV fluorescence. That you can see the shirt glow indicates that the light has been bumped up to the visible range and is not, therefore, UV reflection. A few humans can actually see reflected UV light,[8] but for the remaining 99.999% of humanity—if you can see the object, it is through visible light, not UV.

To add to the confusion, many companies are selling fishing products—lures and lure paints—which they allege reflect UV. The following are some examples:

- A patent application for a fishing product proposes to use Titanium Dioxide—a major component of sunblocks and sunscreens—to reflect UV.[9] Titanium Dioxide absorbs UV, reflecting very little (Appendix B). Nevertheless, this product is being widely marketed and perhaps sold. Here is some of the prolific advertising from the inventor—"The problem then was to find a substance which would be nontoxic yet highly reflective of ultraviolet light. Several crystalline compounds came to mind. For years white pastes of Zinc Oxide or Titanium Dioxide have been safely used by mountain climbers and a few beachgoers to protect their lips, ears and noses from sunburn. These crystals do not absorb ultraviolet light rather protecting against sunburn by *reflecting away the ultraviolet light."* [emphasis mine] (www.salmonuniversity.com/Fool_a_Fish.html)

  What is most stunning about those claims is that Zinc Oxide is even less reflective of UVA than Titanium Dioxide (3-4% vs 6-31% based upon composition) (Appendix B). Both absorb UV.

- Another company is selling a product intended to prevent camouflage clothing from reflecting UV. Their marketing is quite convincing. In their video you can watch the clothing as it glows under the UV light gradually turn darker as their product is sprayed onto it. The visible glow coming from the material is UV fluorescence, not UV reflectance—which is not visible.

- One fishing lure company markets plastic worms and other lures that purportedly reflect UV. Their advertising displays the lures in all the different colors in which they reflect UV—UV green, UV purple, UV blood, UV blue, etc. Now, since humans can't see UV light, the fact that a color is visible is a direct indication that the objects are fluorescing, not reflecting UV.

I mean no disrespect to these entrepreneurs. However they are, at best, misinforming their customers with these statements. The aforementioned companies are speaking of UV *fluorescence* not UV *reflectance*. It was necessary for me to highlight the difference lest the reader misunderstand the intent of this book. This work contains information regarding the ability of trout to see objects through reflected ultraviolet light, not through fluorescence.

As trout fishermen we are not greatly concerned with UV fluorescence; discernible fluorescence does not often occur in trout-stream insects and baitfish. True, some brightly fluorescent lures are quite effective, but they are stimulating other instincts in trout—aggression, curiosity, or territorial behavior—rather than imitating a natural food item.

Throughout the book, the term "VIS" will refer to Visible Light and the term "UVR" will designate UV Reflectance.

So we will focus on UVR in order to improve our imitation of trout foodstuffs. While we cannot see the UV reflected from objects, the camera can and the trout can, and that is what the game is all about.

UVR is of two types:

1. *Speculular reflectance,* as of light bouncing from a mirror. The light—whether VIS or UV—from a single incoming direction is reflected into a single outgoing direction at the same angle to the surface. "The angle of reflection is equal to the angle of incidence." Effectively, the light ray "bounces" from the mirrored surface, as a thrown ball—with no "spin"—bounces from a wall. If you wish to catch that ball or see that light, you must be in its path.

2. *Diffuse reflectance,* a scattered reflection of the UV rays from an object. This is similar to the effect of matte paint on a wall. That you can see the wall indicates that it is reflecting light rays; however, you do not generally see a great difference when viewing it from different angles—it reflects much the same color with the same intensity.

The shiny scales of some freshwater baitfish provide specular reflectance in the UVA. Fly tiers simulate these scales with reflective foil of Mylar or metallic tinsel, which in some cases works to reflect both VIS and UVA. I say "in some cases" because not all metals which reflect VIS also reflect UVA.[10]

Some areas of the baitfishes' body, particularly white underbellies, as well as portions of the bodies of many insects, provide diffuse UV reflectance. Tiers were not previously aware of this, nor capable of determining a suitable substance to mimic this diffuse reflectance—hence this book.

Of note, some fly tying materials, usually synthetic, provide both specular and diffuse UV reflection. This provides a "flash" of UVR depending upon the angle of the light, as well as a consistent UVR tone. This dual presentation is consistent with the scales of some baitfish and the chitin of some nymphs.

I have always rejoiced in the invisible. Certainly I delight in the usual four dimensions, but often some part of me yearns for the simple quiet of dimensions seven or nine, or the bucolic non-spatial pastures of eleven. Frankly, I don't appear to have been designed, either physically or psychologically, for dimensions one through four. This is blindingly obvious when I cast a fly; but it has been evident since childhood from my propensity to walk into stationary objects.

I cannot see ultraviolet light, a disability I openly acknowledge. I was all the more fascinated therefore when I discovered that many creatures, finned or feathered, quadrupedal or bipedal, ran or swam or wriggled or slithered through a world of vision that I would never experience. So, flushed with excitement, I began my research into the world of ultraviolet vision.

I had my "Eureka!" moment, in the stream not the tub, and now wish to share this knowledge with you, the reader. This book may change the way you fish, but I hope it also changes the way you look at the world around you. If you accept one invisible world—the world of ultraviolet light—it is the first step toward a different view of the universe we share.

- Overmywaders

What does an object look like through UVR?

Before further examination of the trout's vision, it might be helpful to appreciate the immensity of the visual perspectives that we, as humans, cannot see. As humans we see the world through trichromatic vision, using three cones in the retina of the eye. These cones send signals to the brain to represent the colors Red, Green, and Blue (RGB), and their millions of combinations.[6] Until relatively recently, scientists thought that most of the animal world had equal or lesser ocular capability—either within man's visible color spectrum or in black and white. Ah, hubris.[7]

To the right is the common Brown-eyed Susan as we see it. Below is a photo of the same flowers with the camera recording only reflected UVA wavelengths of light.

You will note that the center of the flower is darker than the outer part of the petals. This is a result of the outer half of the petals reflecting much of the UV light, while the center of the petals is reflecting very little. The very center of each flower is light—high UVR—against a dark background—UV absorbing. This complete pattern, known as a "bullseye," draws pollinating insects with UV vision, for example, bees and butterflies, to the very center of the flower where the pollen lies.

The bee doesn't only see the UV "color" but some of the same visible color which we see. What the mixture would look like to the bee—or a trout—is something that humans may never know. We can, however, attempt to grasp an understanding of the complexity of the image by superimposing the UVR photo

Bees were said by the classical writers to be both intelligent and hardworking. That obvious oxymoron is made apparent by the fact that bees seem to require plants to inform them of their own business—the collecting of pollen. The "bullseye" is evidence enough of the low intelligence of the average bee, but if you look closely you can almost read the message "Employees must wash legs after pollinating."

- Overmywaders

over the visible light photo. One possible result would be the image below:

Low UVR is important, too

Of equal importance with UV diffuse reflection is UV absorption—the property of minimal reflectance. Some of a trout's food supply has scant UVR, that is, most of the UV rays are absorbed. In that case the prey will stand out starkly against a backdrop of UV scattering substances in the water. The old adage "Dark day, dark fly" was an unconscious acknowledgement of the ability of trout and salmon to see a dark, UV absorbing fly pattern well against a UV reflective background, such as sand and rocks, at a time when VIS is minimal.

Fly fishermen often use black flies at night, though it seems counter-intuitive. The prevailing light at night is UV and, as we shall see through the underwater images in a later chapter, the UV reflected from plankton and suspended inorganic particles in the water produces an ambient UV light. Pass a UV absorbing fly, for example a Black Marabou, through this water and it stands out boldly against the lighter background. A highly UVR fly, like a White Marabou or Chartreuse Marabou, would also attract a fish's eye at night. Interestingly, the actual color would be immaterial as color wouldn't be visible—white and chartreuse marabou would look very similar (see Chapter Eleven).

> **"Some circumstantial evidence is very strong, as when you find a trout in the milk."**
>
> **- Henry David Thoreau**

A few words on UV photography

Ultraviolet reflectance photography has advanced tremendously in recent years. In Appendix A, I explain in some detail the photographic procedures, equipment, and materials I use to produce the photos within this work. However, it is important to understand that the best of UVR photography will not, with our current science, emulate the vision capabilities of a brown trout.

Grayscale (or greyscale) measures the intensity (lightness or value) of the light reflected from an area (dot) of a plane surface and defines each pixel (picture element) as a byte, generally of eight bits, which yields 256 shades of gray. Consider it a limited Black & White photograph. Since only the intensity is measured, two colors, e.g., red and yellow, of the same intensity will appear the same in grayscale.

The trout will be receiving ninety-five percent of the UVA that is striking the water's surface at a ninety degree angle. The lens of the trout's eye will not significantly reduce that amount. The camera, on the other hand, must use a filter to remove all visible and infrared light. That filter absorbs twenty percent of the UVA. Each successive piece of glass in the camera lens takes another thirteen percent. Finally, the digital camera sensor does not respond to all of the UVA input. So, as you look at the UVR photos throughout this book, understand that to a trout the image would be much, much brighter.

Because UV light is not visible, representation of it may be rendered in many colors and styles. Sometimes I have chosen to cast the UV in grayscale—all colors reduced to their value on a scale of white to black. In other instances I have used strong pseudo-colors to accent the difference between the VIS and UVR. And, in other photos I have overlaid the UVR on the VIS as a filter mask in order to more clearly represent the blending of the two as a trout might—just might—see the image.

In all of the UVR photographs, light areas indicate UV reflection—the closer to white, the greater the UVR. Similarly, the darker areas indicate low UVR or UVA absorption—two ways of saying the same thing.

In the following images of artificial flies and fly-tying materials, you will notice the frequent occurrence of a plastic case containing white powders. These chemicals reflect UVA in a predictable manner and provide a means of measuring the percentage of UVR. The chemicals, from left to right, reflect the following percentages of UVA: 81%, 87%, 3%, and 31%. Below they are shown first in VIS, then in UVR.

Where the test chemicals are not present, it is because they were clipped from the finished photo for cosmetic reasons. Sometimes a white and a dark blue/black rectangle are inserted in the image to retain color value perspective.

The fly tying materials shown in this book were provided by Hareline Dubbin, Inc. (www.hareline.com). Hareline's generous assistance made it possible to photograph materials that are consistent in UVR, as Hareline creates and dyes its products using industry standard batch-processing procedures and materials.

CHAPTER TWO

WHY WEDDING PHOTOS ARE NEVER TAKEN IN UV

The resolving power of UV vision is remarkable. UVA is absorbed by some compounds and reflects well from others; the contrast often revealing hidden features of an object. This is why UVR photography forms an integral part of forensic science and NDT (Non-Destructive Testing); cracks, scorched areas, and repairs are often immediately apparent through UVR.

This Pre-Columbian head (above) was repaired carefully by experts after it was accidently dropped. The glue and paintwork that covers the damage contains organic molecules that absorb UV. The image on the left is with visible light, the image on the right is from reflected ultraviolet light. Note how the UVR shows the hidden cracks. (Photo courtesy of Dr. Austin Richards, Oculus Photonics)

Below is a normal, freshly-washed hand photographed in VIS. No scars or other imperfections are apparent.

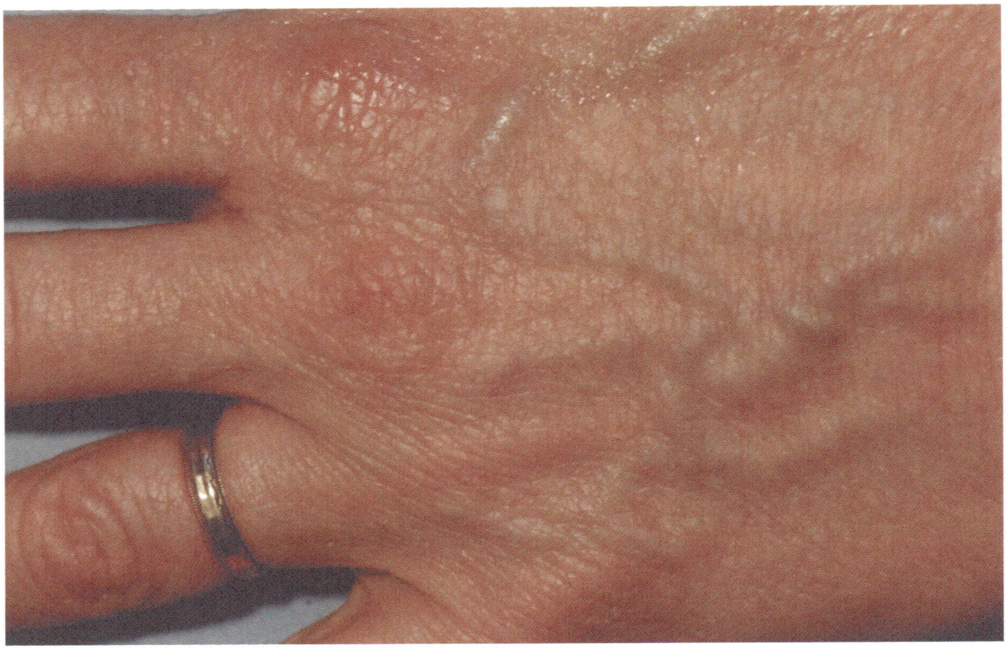

And now, the same hand photographed in UVA in order to record the UVR.

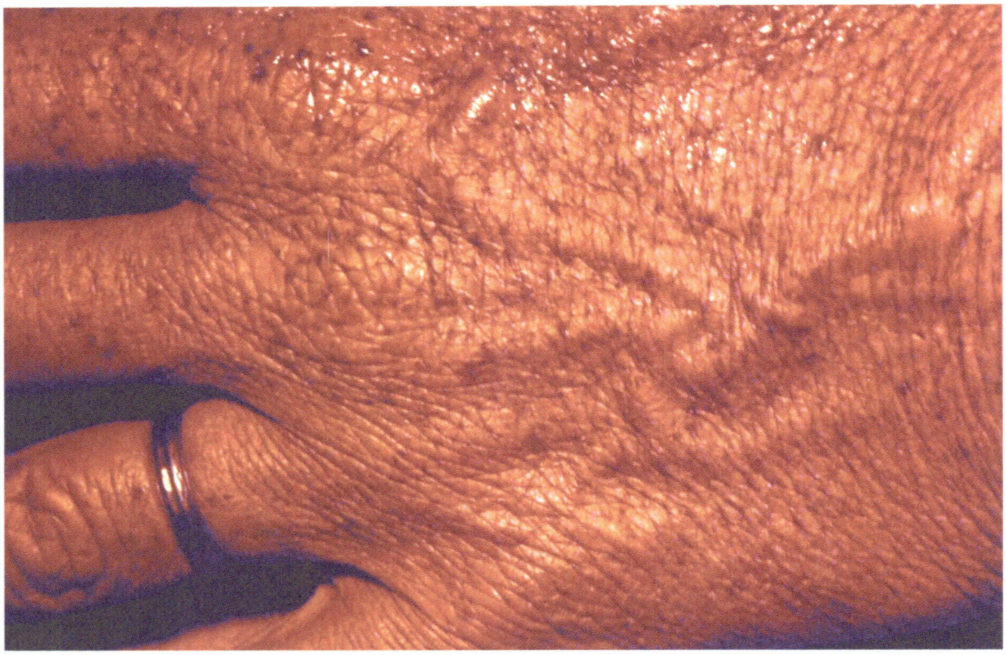

You will notice the dark spots in the upper left, tiny clusters of melanin that absorb UVA, which are visible only through ultraviolet reflective photography. Because melanin is deposited around the site of bruises and bites, forensic scientists are able to detect these injuries up to five months after the event.[11] Notice also the lines and tiny scars not evident in VIS.

There is a fishing lesson here. You may recall that in the late 1950s a line of fly nymphs were produced in realistic rubber compounds with excellent anatomical detail and color. At first these rubber flies sold well, due to their apparent realism. It became quickly evident though, that the only parties who didn't bite on these pretenders were the fish. I never met anyone who caught a trout on these soft, delectable-looking lures. Each lure was the right color, size, shape ... what did they lack?

They lacked depth associated with multiple inner surfaces and the irregular textures common to all living things. Through UVR in combination with VIS, trout have an opportunity to see fine details of the chitin, the outer surface of an insect's body and wings. How deep this vision goes depends, of course, on the individual trout, the conditions, and the insect. One thing is evident: The rubber of the faux nymph was too smooth and homogenous. The dubbed bodies of artificial flies, on the other hand, provide a broken, irregularly-textured silhouette and depth; not unlike the human hand above in UVR.

So, surface texture is significant because, despite what we see with our more limited vision, the trout can detect in the UV that natural flies are not perfectly smooth. For example, we could tie caddis flies with very

smooth plastic tent wings and might catch an occasional fish. However, caddis fly wings are hairy, a "hair" that is visible to the eye along the edges of the wing. This hair-like texture can be simulated with feathers, as in the Henryville Special, or animal hair, as in a sparsely tied Elk-hair Caddis. Neither feathers nor animal hairs are perfectly smooth and this fact is evident both in VIS and UVR. Above is a close-up of the wing of a Henryville Special in VIS. Note the texture of the feather, especially along the edge.

CHAPTER THREE

A MOMENT OF SCIENCE...

If you do not need to know the Why of things, please advance to Chapter Four.

Are you still with me? Okay, now let us examine, with all our logic circuits ablaze, the rationale for trout to require vision in the ultraviolet. Once we understand why trout need UV vision we will be better able to understand how we may apply that knowledge in our art of angling.

First, as a given, trout depend upon vision to capture most of their prey. Certainly the other senses, especially smell, play an important role in the occasional feast; however, trout of the rivers and streams generally find their next meal by observing it drifting toward them, or fleeing from them. Given then that vision is so important for the salmonids, the scope of that vision should be as inclusive as possible to optimize feeding opportunities.

Let us postulate that evolution within a species only retains those functions that are essential to continuation of the species. "Essential" can be something as simple as a colored patch on the cheek of a male bird; if that is the only distinguishing mark to differentiate it from a similar bird of another species. By retaining that color patch, females are able to identify appropriate mates. Now suppose that the bird lives in a forest environment where a certain amount of camouflage is helpful. It would not do to advertise one's presence through gaudy patches of color that predators might note, so, over a period of many lifetimes, the identifying patch is found in the ultraviolet, detectable by other birds but not by four-footed predators. It is worthy of note that all avian families exhibit some ultraviolet reflective feather markings.[14]

Having established that ultraviolet markings have an evolutionary significance, we next determine whether the fishes enjoy the same optical abilities for the same reasons. Birds use UV vision for detecting mates, food, prey, and

The term "evolution" in this work refers to alteration of form or appearance *within* a species and carries no dogmatic baggage.

danger. Insects do as well. What advantage then is UV vision to trout? Well, in the fry stage, trout feed upon microscopic plankton which reflects UV—for this UV vision is convenient. As they grow, trout require more robust food. It is at this stage that many of their UV-sensitive cones convert to blue-sensitive. Naturally, biologists on seeing this diminution of cones leapt to the conclusions that

1. the reduction of UV-sensitive cones occurred across the entire retina leaving adult trout with no UV vision
2. adult salmonids do not require UV vision

We can hardly fault the biologists for these assumptions. Generally, the studies into such arcane subjects as vision in fishes is pursued as pure science with little funding from sources outside the academic setting. Publish-or-Perish is still the rule for many academics. This creates a "get there firstest with the mostest" drive; in other words, a race to publish. Thus, the birth of assumptions. The problem then becomes the prolific citing of that research which eventually sets the assumption as a fact. It may take decades before further research is undertaken that alters/removes an assumption in pure science.

Later studies, alas much later, revealed that UV-sensitive cones are indeed found in adult trout. These cones, however, were not concentrated in the same section of the retina in which they were prolific in the pre-smoltification stages; some consolation, I hope, to the original researchers. The cones in adult salmonids are present in the dorsal-temporal quadrant representing a large proportion (15-20%) of the surface area of the adult retina.[15] Further, it has been determined that the trout retina contains a concentration of stem cells; enabling the regeneration of damaged cones and rods.[16][17]

The question for anglers is "Why do mature trout have UV vision?" or, more properly, "Why do mature riverine trout have UV vision?," since UV vision seems entirely appropriate for deep lake dwelling species functioning below the level of visible light.

This brings us back to the uses of UV vision—detecting food, prey, mates, and danger.

Food/Prey:

If the primary form of food for the trout after smoltification is aquatic insects, then any optical ability that enhances seeing/identifying aquatic insects has positive evolutionary value. Just as anglers often wear polarized sunglasses to cut glare from the water and see more clearly, trout use polarized vision for improved sight. With trout, the polarized vision is concentrated in the UV-sensitive cones, enabling them to see greater detail in adverse light conditions.[40]

Much of the significant aquatic insect activity occurs at dusk or later. While the rods of the trout's eye are responsible for some of the sensory input during the period of dusk through dawn, rods cannot provide detail—that is the function of cones. The ultraviolet sensitive cones permit trout to see fine details during these low light periods. This ability of both cones and rods to function at the same time is known as mesopic vision—that phase between "photopic" vision—cones-only daylight vision, and "scotopic" vision—rods-only low-visible light vision.

Since the rods of the trout's eye are adequate

in providing gross vision, but not in capturing detail, how does a trout feed after dusk? As most fly fishermen know, some of the biggest trout only start their feeding at dusk and continue into the night. Many of us have stood in the river, uncertain where the shore lies but entirely sure that the sound as of a cannonball being dropped in the water twenty feet away was the slurp of a large trout taking a Hex fly. But how does the trout know it was a fly and not a twig, if no detail is available? I will posit that the UV sensitive cones are still providing detail, because the UVA is still present. Moonlight and starlight don't provide as much UVA as the sun by day, but there is still sufficient to be captured by the sensitive cones of the trout's retina and interpreted by his brain.

Cloudy, turbid water conditions, perhaps caused by rain washing fresh sand or silt into the stream, can provide an excellent feeding opportunity for trout, while at the same time greatly restricting the trout's VIS vision. The quality of the trout's UV vision in turbid water depends on the type of sediment which is present. Fine sand will scatter UV rays, thus providing a good background for prey items; black silt will absorb UV as will yellow water, thus reducing the UVR available to the trout. However, UVR generally improves the fish's chance of capturing food by distinguishing the food items—which might be brown nymphs—from similarly VIS-toned brown silt.

The images to the right illustrate how UVR allows fish to see objects clearly against a background of ambient UVR caused by UV scattering from plankton. This would be true in either fresh or salt water environments as both contain abundant plankton. (Photos courtesy of Professor Thomas Cronin, UMBC)

For surface feeding during the daylight hours, ultraviolet vision makes it possible for the trout to distinguish between an insect and a twig—or an insect and an artificial fly. Your average small twig from a tree will have an overall low UVR. An insect, however, will usually display many different intensities of UVR. An artificial fly may, or may not, have various UVR markings, and such markings may not accord with the insects currently on the current. On the other hand, the white paper of a filter-cigarette butt shines brightly in the UVA, and the filter, if brown, will show some interesting UV markings. Have you ever wondered why even a wily old trout will sometimes rise to examine a cigarette butt?

Mates:

This is mates as in mating, not Australian male bonding. Do female trout chase off the males which exhibit low UVR markings and only mate with those adorned with the strongest UVR? This is common with other fish species, e.g., sticklebacks, as brightness of UVR markings is often a reliable indication of health and status.[18]

Danger:

UV vision permits greater discrimination between objects. A stone projecting out of water may be a light tan in color and have a medium UVR, while the man standing on the stone may be wearing clothing also tan in color but with a high UVR (see Chapter Thirteen). Those who stalk wary trout are all too familiar with performing the most secretive approach to a pool, only to watch a nice trout bolt for cover. Adequate camouflage must include camouflage in the ultraviolet.

It would seem we have answered the question "Why do trout have ultraviolet sensitive vision?"

CHAPTER FOUR

HOW SOME ANIMALS USE UVR

In order to more fully appreciate UV vision, let's look at some of the ways it is used in the animal world.

Many birds have excellent vision in the near-UV. A small European falcon, the Kestrel, provides a startling example of this. The kestrel feeds primarily on field voles, *Microtus agrestis,* a four-inch rodent with the habit of marking its path through high grass to and from its nest with spots of urine. The urine's high UVR provides the vole with visible trail blazes for its return to the nest. Unfortunately for the vole, a hungry kestrel sixty feet above is also able to see these bright spots of UVR. The circling kestrel can detect the path of the vole across a dense meadow, determine the freshest path by the higher UVR of the newest urine patches, and track the rodent to its current location. Dinner is served.

The issue of incontinence among small rodents has never been properly addressed by society. Efforts should be made to ameliorate this ongoing death-by-urine. Doubtless many of these voles would have survived to return to their families rejoicing; if only they had been properly diapered.

- Overmywaders

Some songbirds, visibly identical in VIS, use UV patterns for mate selection. This is especially convenient for ensuring integrity of similar subspecies. Another avian use for UV vision is "Honesty markings." These are UV-reflective marks on the male birds which deteriorate with age. The discerning female of the species uses these markings to determine the courting bird's age and vitality. Thus, though the strutting and posturing male may act young, the female can detect, through UV vision, her wooer's actual age and make her choice based upon that.

I see a ready market for Avian Grecian Formula.[19]

- Overmywaders

Interactions between competing male birds of some species use the UV honesty markings as well. A vigorous male will pay little attention to another male whose UVR is low and will reserve his energy in mock battles for a more viral opponent.[20] Among insects, the females of certain species of butterflies choose their mate based upon the size and brilliance of UVR wing markings.[21]

Whitetail deer browse during dusk and dawn, the periods of day known as crepuscular. To aid them in this, whitetail deer do not have a UVA filter in the lens of their eyes. Thus they are able to see approaching predators in even the very dimmest of light. While there is only little light at night, what light is available is shifted towards UVA-wavelengths.[44][45]

Mayflies, lacking means of attracting their mates by vocalization (ah, the dulcet tones of the mating cry of the March Brown...), use a pattern of UV visible markings instead. The mayfly has vision in the UV range. This allows them to find mates in the hours of dusk when the prevailing illumination is ultraviolet. For the swarming mayfly, species and gender specific markings will be UV reflective, easily seen by potential mates, though not discernible to the human eye. Their life is brief enough—what mayfly wants to spend his/her last moments mistakenly courting a member of the wrong species or gender?

A wet highway is often covered with mayflies in their mating dance because the wet asphalt has the same general UVR as a river.

The razorback sucker has a fascinating use of UVR. A male razorback, upon detecting another male within his territory during spawning time, will roll his eyes, much as a teenager does to the pronouncements of a parent. The eye roll sends a flash of UV at the intruder, who typically responds by leaving the area.[22]

CHAPTER FIVE

UVA IN WATER

Ultraviolet rays in the UVA range penetrate fresh water with ease. When entering at a 90 degree angle only 5% of the UVA is reflected—the rest passes into the water (see Appendix B). This permits trout to see their prey even at depths, or in conditions, which filter out most visible light.

Visible light is lost according to its wavelength, with the longest wavelength, red light, ending at the shallowest depth—usually ten feet or less depending on the clarity of the water. As you go progressively deeper, you lose most of the orange by twenty-five feet, then yellow, and so on, down through the spectrum.

UVA wavelengths generally penetrate to depths of one hundred-sixty feet.[23][24] Thus, fish are able to see objects in the UVA, either through UV reflectance or UV absorption, at greater depths or in more turbid water than VIS.

Recent work using submersibles found amounts of UVA required by some deep-sea dwelling crustaceans is available at 1800 feet beneath the surface.

The photos on the following page represent identical underwater views taken with green (top) and UV (bottom) filters. Note the enhanced contrast between fish and the background in the UV-filtered image. The images below are equally striking.

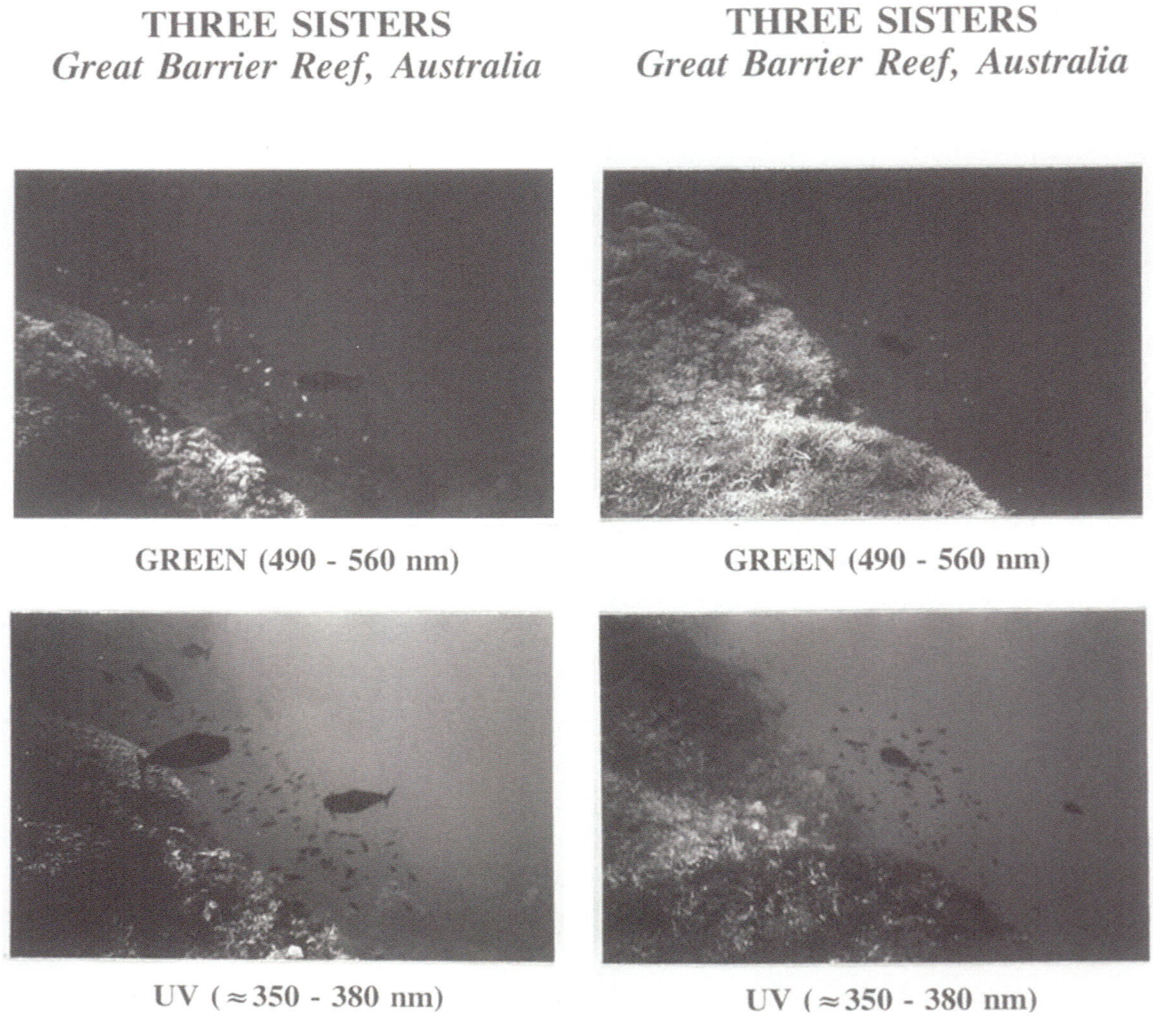

Photos courtesy of Professor Thomas Cronin, UMBC

MACGILLIVRAY'S REEF
Lizard Island, Australia

GREEN (490 - 560 nm)

UV (≈350 - 380 nm)

MACGILLIVRAY'S REEF
Lizard Island, Australia

GREEN (490 - 560 nm)

UV (≈350 - 380 nm)

The following images illustrate the percentage of UVA that penetrates the surface of water. Most of the water in the UVR image is dark; this indicates that it reflects little UV—the lighter the tone, the greater the UVR. The lighter shade of the current flow midstream is a factor of the time lapse photography required by UVR. The camera converted bits of UV reflective foam into light lines as the foam moved downstream. The granite bridge in the UV image is highly UVR, while the plants by the river are medium UVR, and the climbing ivy is low UVR.

Zooming in closer we can see that each type of plant has a different UVR signature.

Notice also that the UV doesn't reflect back to us from under water. This is evident when we look at the plants near the bank (white arrows) that are underwater in the VIS photo, but don't appear in the UVR image.

Trout are able to see polarized light—but only in the UV. Of what use is polarized UV light? Well, bees use polarized UV light for navigation, perhaps trout do as well.[25] However, the immediate point of interest to anglers is that the trout, protected by polarization, will not receive UV glare. The clarity of the trout's vision in the UV will be unimpeded—which cannot be said for the trout's color vision.

From the trout's perspective, the UV is providing a well-lit ambiance. Low lights, quiet music, and the reflection of candle flames flickering in crystal water glasses may make our dining experience enjoyable—though I'd rather see what I'm eating—but for the trout, capturing his meal must occur before fluffing his napkin. Suspended particulates in the water, both organic and inorganic, are reflecting UV; this provides a backdrop for any drifting or swimming matter. Even on a cloudy day, a small fish or nymph drifting on the current, backlit by the ubiquitous UVR, would be readily apparent to a hungry trout in his feeding zone. Our object therefore, is to create imitations which are not only visible on this stage, but which provide the UVR that the trout expects from his dinner.

CHAPTER SIX

IN THE TWILIGHT

For the trout fisherman on running water, two times of day stand out, dawn and dusk. Sure, many fishermen pound the water all day long as catharsis, exercise, or simply for the joy of it; but in many rivers the morning and evening are the best times to fish. Some, like me, who know they have only so many casts per day before they require oxygen, save their energy for the evening rise.

"The evening rise" is a phrase familiar to anyone who has ever spent time sitting on the porch of a cabin or around a campfire near a good trout stream. Summer dusk on a river is an almost tangible memory to carry us through a long winter.

Large brown trout commonly feed in the evening. The upper stretch of the Manistee River in Michigan provides a good example of this. During June you can wade the cold water in the hot sun all day and see no fish on the smooth, sandy bottom. The fish are there, but they are small brook trout very light in color that blend well with the streambed. The large trout—browns and rainbows—spend the day sheltered under fallen logs or man-made structures built against the undercut banks. As twilight comes on, the tiddlers discreetly disappear from sight and the monsters come out to feed.

The light is poor for the fisherman at dusk. The cones of his eyes no longer send colors and details; the more photon-sensitive rods of the eye must now provide what little vision he has. The trout, on the other hand, using the dominant ultraviolet light can see the details of their submarine world just fine; both the rods and the UV sensitive cones are in use.

The cover of darkness provides trout some security from many of the predators—kingfishers, heron, and mink—that they fear by daylight. The mayflies are coming out, too, either hatching into duns (subimagoes) or as adults (imagoes) looking forward to their mating dance and dressed accordingly.

Now is the best of the fishing; the angler has enough light to see a rise near his fly, the insects are active, and the trout are eager to feed. Now is also the time when UVR is most important for all concerned—man, insect, and fish.

If there isn't any insect activity, the large trout may move into the shallows and chase minnows. This continues well into what we know as darkness, and on those waters where trout fishing is allowed at night, the fishermen pull out their large streamer patterns or a deer-hair mouse, and cast hopefully into the quiet black that surrounds them. The trout, meanwhile, are cruising comfortably in waters lit by ultraviolet light. Now is the angler's opportunity to use a Black Marabou streamer, or a White Marabou streamer; something with maximum contrast in the UVR. Night-time trout are not fussy about imitation; they are hungry and feeling secure under the cover of darkness.

Dawn and dusk, or day and night, have been defined in similar ways long before we had clocks. The demarcation between day and night was based upon the loss of color vision, that is, when the cones of the retina are no longer effective and the rods take over. The rods of the human eye are unable to distinguish fine detail or color. Similarly, day was signaled by the ability to discern colors.

Hence in ancient Israel:

> **"The undisputed terminus a quo [earliest time] is given as the moment in the morning after dawn when it is first possible by natural light to distinguish between blue and white. In mishnaic times the tzitzit [tassels, fringes] of the tallit [prayer-shawl] still had one blue thread among the white threads."**

Berakhot, Chapter One, Mishnah Two

And from the Koran:

> **"...And eat and drink until the white thread becomes distinct to you from the black thread of the dawn..."**

[Al-Qur'an 2:187]

Dawn provides much of the same. Sometimes the angler takes advantage of rises to early insect activity; otherwise he may find casting small streamers effective. The large trout are looking for their last charge of protein before heading to a deep hole behind a boulder or some other safe holding place for the balance of the day.

There are, of course, many rivers that offer an opportunity to catch large feeding trout throughout the day. Spring creeks provide a fairly consistent water temperature, as do bottom-flow dam tailwaters. In both these environments trout may feed eagerly at any hour, especially if they have the nearby shelter of floating cress or thick aquatic plants. Insect life flourishes in such water and the trout grow large quickly.

For those of us who fish freestone streams it is different. These streams and rivers often have little protection from the sun, having lost the cover of trees, either through tree-cutting or the meandering of the stream. The water temperature fluctuates along with the level of the aquifer, the air temperature, recent rains, the dominant color of the stream bottom, the water depth, nearby roads and parking lot run-off, and more. Thus the insect life is not as abundant as that which a more stable water flow and water temperature could support. The larger trout avoid the mid-day heat and snooze the day away until the first shadows strike the water. The insects also wait for the light to dim, though for different reasons. For some aquatic insects, the larger they grow in the nymph stage, the more certain it is that they will do their drifting downstream after daylight hours to avoid the numerous small fish which prey upon them by day.[26] Other aquatic insects complete their emergence into the dun stage during twilight or later in order to avoid predators as well as the deadly desiccating heat of day;[27][38] while yet another factor is the greater ease of detecting appropriate water-bodies when the sun is lower in the sky and polarized light is dominant.[38]

Large trout are great Noel Coward devotees:

"Mad Dogs and Englishmen, go out in the midday sun.
The smallest Malay rabbit deplores this stupid habit.
In Hong Kong, they strike a gong, and fire off a noonday gun.
To reprimand each inmate, who's in late.
In the mangrove swamps where the python romps
there is peace from twelve till two.
Even caribous lie down and snooze, for there's nothing else to do.
In Bengal, to move at all, is seldom if ever done,
But mad dogs and Englishmen go out in the midday sun."

- Overmywaders

CHAPTER SEVEN

MAYBE IS SOMETIMES AS GOOD AS YES

Fly fishermen always enjoy a pleasant protracted discussion of the relative merit of presentation, size, color, and outline in enticing trout with a fly. You've listened to these debates...

This is the never-ending argument and it won't be resolved in this book... because we are going to adroitly side-step the issue. All of those factors are important in successful fly fishing for trout. However, ultraviolet reflection adds another category altogether. If all of the aforementioned factors are correct, what element of the fly tips the scale for the trout, inspiring the acceptance of the artificial fly as food? That would be appropriate UVR. The trout know what they want and we *must* give it to them.

What do the great prospecting or searching patterns of fly offer the fish? Let's examine a few of the usual prospecting patterns.

First, the Hairwing Royal Coachman. The Hairwing Royal Coachman and the similar Royal Wulff resemble no known insect, yet trout will rise eagerly to them. Since a hatch of Royal Coachmen has never occurred, size, color, and outline are moot as standards. Proper presentation is, of course, still important. However, what drives the trout to take the fly is the marking that indicates a natural insect—the wings have a very strong UVR signature.

In all the UVR photographs which follow, the light areas of a fly or material indicate UV reflection, the darkest areas indicate UV absorption. The closer to white, the greater the UVR.

The Fanwing Royal Coachman, a very killing fly which was largely replaced by the more durable Hairwing Royal Coachman, has an even more remarkable UVR signature.

Either of these flies dancing over the riffles of a stream in bright sunlight would capture the attention of a trout. However, so would the same fly with yellow wings. What the yellow wings would not typically provide, and the white do, is the strong UVR marking. And a trout upon seeing a Royal Coachman settling on the water at dusk, when colors can no longer be clearly distinguished, would immediately notice the strong UVR of the wings in the prevailing ultraviolet light.

Another fly that is often used in prospecting for trout is the Wright's Royal. The Wright's Royal, for those unfamiliar with it, is the tailless down-wing version of the Hairwing Royal Coachman above. The Wright's shares the strong UVR properties of the rest of the Royal clan but, like the others, resembles no known insect. Yet the fish take it eagerly. As anglers we often use the Wright's on rivers which have a predominance of caddis flies, though no caddis ever looked like that.

The Stimulator, as the name implies, is intended to stimulate the trout to rise. Stimulate it does, simulate it doesn't. I have heard many suggestions of insects that the Stimulator faithfully portrays, the most likely being a stonefly. Yet trout take a Stimulator on water that sees few stoneflies, so there must be something to the pattern other than mimicry. With that buggy-looking hackle and hair the Stimulator screams "bug." The high UVR of the wing, body, and tail; along with the medium UVR of the head, further enhance the impression of life.

Not Vision, Perception

Any trout fisherman, especially those who fish with dry flies, will tell you that trout possess acute vision. Scientists will agree. How is it then that a large brown trout will see this –

RED QUILL

and accept it as this:

Ephemerella subvaria (M)

(Photos from *Streamside Guide to Naturals and their Imitations* by Art Flick)

The answer lies not with the trout's vision, but with his perception; that is, how the trout's brain registers the sensory input of vision and matches it to an established pattern. As fly fishermen we have for centuries looked at the thick hackle of a standard upright dry fly and assumed that the trout sees the hundreds of light points breaking the surface as the six legs of a mayfly. That astounds me! A fish with such fine vision will see all the hundreds of those discrete hackle points. However, while the trout can *see* the difference between six sparkling points and hundreds, the trout can't discern or *perceive* the difference.

Vision is the portal for raw sensory information. That information is so complete, and flowing at such a rate, as to be overwhelming. No animal could survive if each day it had to relearn, moment-by-moment, that certain images represented danger and others, food.

So, cognitive systems, both machine and organic, use pattern-matching. "Pattern-matching" is the act of checking for the presence of *all* the elements of a given pattern. As humans we unconsciously learn pattern-matching at an early age. Since we humans serve as both predator and prey, our brains need to be able to alert us to danger, or food, quickly. This cannot be a conscious process as that would be far too slow; thus, the pattern-matching. When my vision sends images of, for example, a three-legged chair, to the brain, the brain temporarily records the input, while trying to create an object from all the associated parts. This becomes the *gestalt*—an organized whole that is perceived as a whole—rather than a collection of discrete elements.

Having established a *gestalt*, an apprehension of an object, the brain must now match it against familiar patterns that indicate danger in order to send a fight or flight warning. The chair is quickly noted as not matching the shape of a known threat, nor as food. The brain next performs a lower level of mapping which covers more databases as it were.

It is known that humans have a database for tools and common objects—the first database that starts to fail us with age. Sigh.

Eventually—in real time it may be thousandths of a second—the object is perceived as a chair. Important note, it is *perceived* as a chair, it is not *seen* as a chair.

The trouts, fellow predator and prey species with humans, also use pattern-matching for quick identification of danger and food. This mechanism has been proven to be active in other fish[12] as well as many insects.

As the Quill Gordon floats within the trout's range of vision—and here I am going to avoid the complex issues of Snel's Circle, reflection, and refraction and simply assume that the visual sensory input is very detailed and complete—the trout's brain receives input of the fly exactly as it appears from below, in the full trout spectrum, VIS and UVR. The trout brain now gathers the elements that are attached to each other—hackle, body, wings, tail—ignoring floating particles of foam nearby, and assumes that it forms a whole unit. Against this *gestalt* the trout brain uses pattern-matching, just as we would. The order of conditions is presumably the same:

- First, check for danger.[13] Is the object a known threat? "No."
- Next, check for food. Is the object a food item? "Yes," "No," or "Maybe."

And that is the crux of it. If as anglers we can establish "Maybe," we have won the first part of the game. "Maybe" can indicate insufficient information which may lead to further investigation through other trout senses—Taste and Touch. In order for the fish to touch and taste, he takes an object in his mouth, hands being in short supply. This is why we set the hook on a rise, for we know that the trout will reject our fly as soon as it fails the taste/touch test. In short, we have not fooled any of the trout's senses; we have, however, baffled the visual pattern-matching. And if that trout is released or gets free, it has a new template to add to the database of known dangers—a #14 Quill Gordon.

As fly fishermen we learn that during a steady hatch of flies, trout will not take every natural fly that drifts within reach. Some they may follow before taking, others they ignore altogether. This is because the food is flowing in such abundance that the trout can afford to take those that match most accurately one particular pattern.

One instance of this would be when two hatches are occurring simultaneously, a small mayfly and a large mayfly. If the small mayfly started hatching first, the trout would already have established a perceptual pattern-matching that worked for that fly. Therefore, and perhaps we have all seen this, the trout continue to take the small flies and ignore the larger flies. As humans this seems illogical, there is a greater reward of protein and fat in the larger flies for the same amount of effort. Sipping the smaller flies is inefficient. However, let's look at it from the perspective of a fellow predator/prey species—the safe known is always preferred to the unknown.

Suppose I walk outside my cave, club in hand, eager for food and see a small rabbit, whose cousins I have eaten many times. At the same moment, I see nearby a much larger animal that is unknown to me. I do not know the larger animal as a threat, but neither do I know it as food; it might be poisonous to eat. If I throw my club at the larger animal and do not kill it, it might kill me. On the other hand, the rabbit is delicious and not dangerous. Decision made; accept the known pattern.

> **If you are reading this it is probably because your ancestors made the safe choice and enjoyed hasenpfeffer that night.**
>
> **- Overmywaders**

In case any reader thinks the previous paragraphs have been attributing some form of logical thinking to trout—clever little beasties—that is not the case. Cleverness and rational thinking are not synonymous; a moment's observation of CSPAN proves this. To be clever is to be "adept at seizing" from which the word is derived. A trout that is not adept at seizing food when it presents itself is a very thin trout.

Pattern-matching is not a conscious thought process. It is a simple, spontaneous database query, or series of queries, of which the trout—or man—is largely unaware. For example, when you see the word "food" you are pattern-matching. Your vision provides the input of the separate symbols of the letters, but your perception of the letters is a pattern. This required no conscious thought from you. You did not consciously decide to understand the discrete symbols (letters) as a single pattern, in fact you had no control over it. Further, you would have perceived the same matched pattern in many different fonts, despite the visual dissimilarities. You will use the same pattern for **Food** *Food* **Food** *Food* FOOD and **Food**.

For the trout, the outcome or result of the pattern-matching is a "Yes," "No," or "Maybe" that provokes action. "Yes" has the trout take the fly, "No" has the trout reject the fly, and "Maybe" provides different responses for each fish or species. A brook trout, which as a species are opportunistic feeders, will probably respond to "Maybe" by taking the fly in its mouth for further testing. Some large brown and rainbow trout on a quiet stream might drift behind the fly for some time, for more visual sensory input on a "Maybe;" while other trout in faster water might immediately perform the taste/touch test.

Now let us speculate about the elements or parts of a pattern that would be essential to the trout in order for the fish to perceive a floating mayfly as potential food. This would be the trout's pattern—the term "template" is also apt—of a mayfly, without which the fish could not safely feed. The trout's *gestalt* would be the complete image of the floating object as received from the senses. The pattern-matching would be application of the *gestalt* against one or more patterns.

What works in our favor is that the patterns that the trout have for mayflies, caddis flies, etc. are necessarily vague. In order to understand how complex pattern-matching is for the trout, we can observe the stream after a hatch of mayflies. Some of the flies we see will have no wings emerging, they are still stuck in the nymphal case; others will have broken or bent wings; some will have malformed wings; some will be tipped on their sides; some will be floating drowned; and some will be spinners with wings outstretched. The trout's pattern for mayfly wings, therefore, must be quite vague, perhaps simply a small extension from the body, light in color and displaying a touch of UVR. A trout that only eats mayflies with perfectly formed wings is missing a lot of food. The same applies to tails. A mayfly may have either two or three tails. That is the perfect insect under perfect conditions. However, to optimize feeding opportunities and still retain a measure of safety, the trout brain probably reduces mayfly tails to the lowest common denominator of one bent blur.

Let's look at an example and see how generalized it might be. Here is a view of a mayfly from beneath set in grayscale. I used grayscale because any master pattern for mayflies in general must accommodate many colors.

Now, we can't have that many legs or tails as these may be missing so we generalize a bit more by removing two tails and all legs.

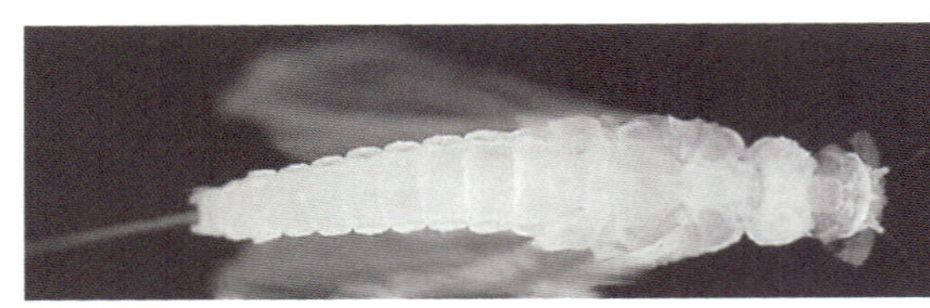

However, as the segmentation of the body is still clear, we need to blur things a bit in order to allow for differing lengths of segments.

That is pretty general. It would even match very well the *gestalt* of a No-Hackle pattern as seen from below. However, the above, besides being pure speculation, represents just one of hundreds of patterns a trout would have for a generic mayfly. Why hundreds of patterns for a generic fly? Because the fly could be viewed from hundreds of angles—front-on, directly underneath but tipped to one side or another, end-on, etc. Below is a directly side-on, or a mayfly lying on its side on the water. I have removed the legs and reduced the tails from three to one.

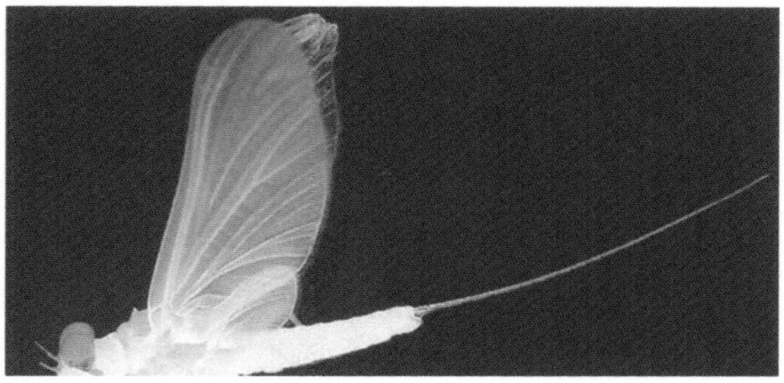

Now I need to generalize it to allow for crumpled wings, broken tails, and differing lengths of segments.

The trout's internal mayfly patterns/templates allow us, through their necessary vagueness, a measure of freedom. But we still need to ask ourselves, "What elements of a natural fly are absolutely essential for the trout brain to use in pattern-matching?"

In the following list of possible mayfly pattern elements we are assuming that no hatch is occurring and the fly is drifting freely; therefore, matching a given fly size and color is not a consideration.

1. Body – The visual sensation should indicate a body, as that is pronounced in a mayfly and provides the greatest caloric offering.

2. Wings – These might or might not exist depending on the state of the emergence, how the fly is aligned, malformed wings, etc. A short post in moderate to high UVR would probably provide enough.

3. Outline – The mayfly outline both from below and from the side is widest near the front, in the thorax. From there it tapers to the tails. An artificial with hackle fore and aft, like the Buzz Hackle, would not present this piece of the pattern. The average twig doesn't, either.

4. Translucence – Many mayfly duns and spinners have translucent bodies.

5. Color – There is no one color that fits as a pattern element; however, there are certainly colors that do not fit, blaze orange for instance.

6. UVR – Some UVR should be expected from any of the mayflies, however the entire fly would not give a consistent UVR. A twig, on the other hand, might give a consistent UVR.

7. Texture – The surface of the mayfly will not be perfectly smooth.

8. Action – Only certain movements would conform to the trout's pattern of most mayfly duns and spinners. In most cases no motion, other than that produced by wind and current, would be correct. Caddis flies would differ, of course.

There are doubtless other elements to a large trout's general mayfly pattern-matching. Each trout would have a unique set of patterns, built through its lifetime exposure to natural mayflies. It is entirely likely that some essential patterns of food sources are passed through genetics as well. After all, trout fry have no parent present to instruct them on the edibility of plankton, nor to transition them to insects when plankton no longer satisfies.

As a hatch is progressing, it is likely that the trout is modifying its base pattern to include the parameters of size and color. The new pattern, through its greater number of elements, increases feeding safety and speed. Thus the trout sips only the small sulphurs which have no trailing shuck while ignoring the larger flies that are hatching simultaneously.

The *gestalt* of the floating fly is going to differ moment-by-moment. Imagine the number of possible views of the fly as it is drifting on the current, perhaps side-on, or spinning slowly in an eddy; and the complexity of the instantaneous pattern-matching required by the trout's brain becomes awe-inspiring. For example, look at your left hand held flat in front of you—you immediately recognize it as a hand. Now, rotate it slowly and observe the different perspectives. In each degree of rotation on each axis the hand has a different appearance; yet your brain still recognizes it as a hand. The trout is going through the same process every time it observes a potential food item above it. As the floating fly turns, a new *gestalt* is formed and the trout's brain must apply a fresh pattern against the new *gestalt*. An utterly unconscious act, yet absolutely amazing!

Remember—all that trout's superb vision simply builds a database query. If the query doesn't return a "Yes," a "Maybe" may be just as good.

CHAPTER EIGHT

BAITFISH AND UV

The term "baitfish" may be misleading when it comes to trout; large trout will snack on any smaller fish regardless of species. Nevertheless, most streams and rivers that support trout also harbor populations of one or more species of small fish such as shiners, chubs, dace, or stickleback. Learning to create imitations of the local fry can bring some large trout to hand.

Even baitfish get the blues, or the ultraviolets. It has been found that some shoaling fish, that is fish that form large groups, choose their shoal based upon what they can see of their prospective shoalmates in the UV.[28] Even the complex maneuvers that schools of small fish perform may be coordinated by mutually observed UV reflectance. So applying UVR materials to streamer flies and bucktails would be entirely appropriate.

Below is a UVR image of a tiger trout fingerling. You'll notice that in the UV, the strong body markings are less apparent, but the lateral line, not seen in VIS, is apparent in UVR. The fins, which are yellow in VIS, are UV absorbing and appear dark; while the throat area is very high UVR.

When tying flies to imitate baitfish certain

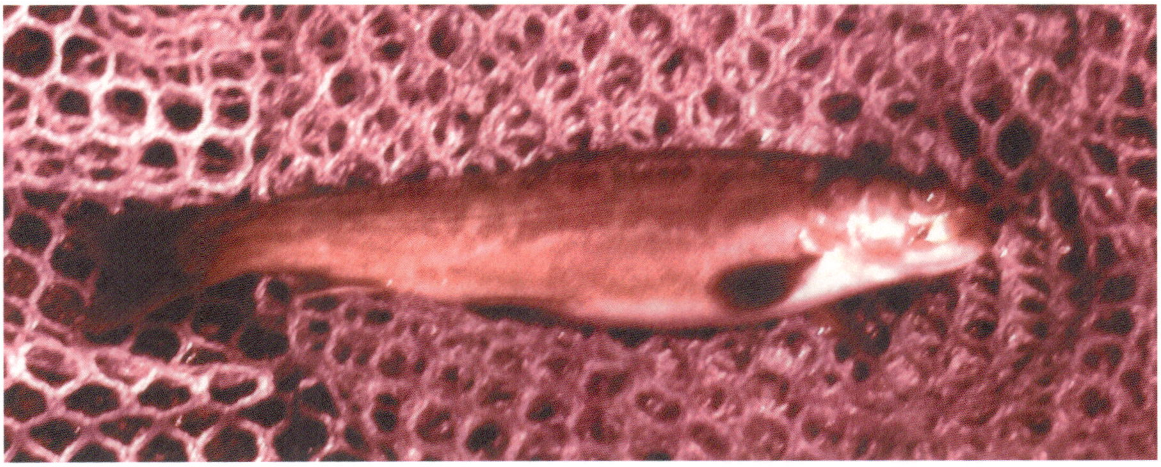

assumptions are fairly safe. Yellow areas will almost always absorb UV, while white areas generally strongly reflect UV. These generalizations are possible because fish predominantly, though not always, display UVR markings through the use of pigments.[42][43] Conversely, the UVR markings on birds are almost always a result of the complex structure of the feather[19] and how it refracts/reflects light. The same structural reflectance is true of many insects, butterflies being the most notable in this regard. Thus among songbirds, the male and female may be identical in VIS coloration, but the male may have brilliant UVR markings where the female has none.

Specular UV reflection is also prominent in some baitfish. Below are some shiners and, true to their name, they shine. In fact, they shine with specular reflection in UV as much as, if not more than, they do in VIS.

You may notice that in VIS the golden shiner above has a light broad stripe running above the lateral line from cheek to tail. In order to highlight this feature I reversed the colors from RGB to RBG in the image below:

The stripe—shown above in purple—does not appear in the UVR images. Another reason to remember that imitation is imitation of color *and* UVR. Below is an indication of the specular reflection of this shiner in UV.

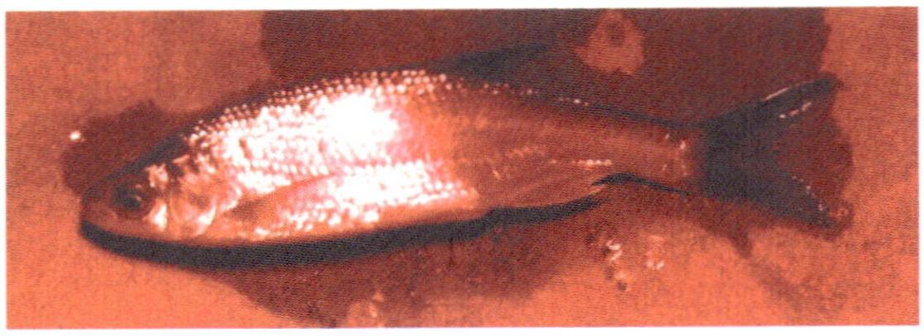

On overcast days or when the sunlight is not direct, that specular reflection will be greatly diminished, whereas diffuse UVR remains undimmed. Below is the diffuse UVR of the shiner:

Note the strong UVR of the belly and flank. The pectoral and anal fins exhibit high UVR, while the dorsal and caudal have low UVR. There is nothing about the dorsal and caudal fins in the VIS to indicate a difference in UVR.

The vast number of species of small fish available to trout in streams, rivers, and lakes across our planet make a full UVR representation impossible. Therefore I will end this chapter with a few cautious recommendations:

- Expect some specular UVR from all large-scaled species of baitfish. If the scales reflect well in VIS, they may in UV.
- Yellow patches, even a faint yellow on fins and tail, probably indicate UV absorbance.
- White patches on cheeks or lower jaw, whether scaled or not, will probably be highly reflective in UV.

CHAPTER NINE

INSECTS AND UV

Aquatic insects of interest to most fly fishermen fall into the "EPT" group, i.e., *Ephemeroptera* (mayflies), *Plecoptera* (stoneflies), and *Trichoptera* (caddisflies). The order *Ephemeroptera* has more than three thousand species worldwide with an estimated six-hundred and thirty in North America. Since the dawn of fly-fishing literature, the mayfly has held a prominent place in our sport.

I was surprised to discover that no entomologists have undertaken research regarding EPT insect markings in the ultraviolet, given the scientific interest in the UVR markings of other insect groups. Yet that is the case. Perhaps the increasing use of ultraviolet reflection photography will invite more interest in the EPT.

Like the UVR of birds' feathers, the ultraviolet reflective markings of mayflies are structural. In mayflies the UVR is reflected by microscopically thin layers of chitin interspersed with airspaces as in the male jumping spider to the right. Visible color has no bearing on the UVR. In fact, most mayfly spinners have transparent wings, yet the markings in the UV are different for each species.

Using light and electron microscopy, they imaged the colorful body scales of males and found that each scale is composed of a sandwich structure with two outer corrugated chitin plates surrounding a thin layer of air (Figure 1C). This structure functions as a multilayer reflector, whereby the alternating layers of chitin, with a high refractive index, and air, with a low refractive index, organized at just the right periodicity with respect to the wavelength of light reflected, result in constructive interference, as light reflected off each interface emerges in phase (Figure 1C). The result is a brilliant iridescent display by each scale, with a large reflectance peak in the human visible range—the color of which changes from green to yellow to orange depending on the viewing angle—and a smaller peak in the ultraviolet (Figure 1B). Interestingly, males also have some body scales with the same chitin plates as described above, but lacking the air gap, which results in a dull purple reflectance (with a relatively large UV contribution).

"Animal Coloration: Sexy Spider Scales"
Lisa A. Taylor and Kevin J. McGraw
***Current Biology* Vol 17 No 15**

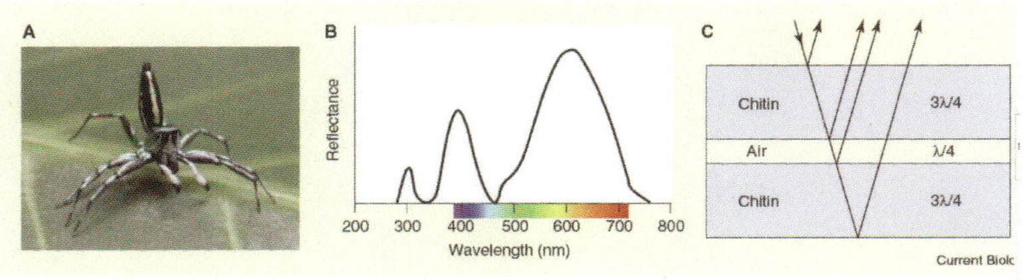

Figure 1. Male coloration in *Cosmophasis umbratica*.

Ephemeroptera – Mayflies

As mentioned earlier, mayflies must use their acute UV vision in order to locate water in the low-light of dusk or dawn for their mating and depositing of eggs. True, they sometimes mistake asphalt roads for rivers, but enough get it right for the genera to carry on. Surprisingly, documented proof that mayflies carry species-specific and gender-specific UVR markings in order to facilitate mating has not, until now, been available. The images which follow represent not only the first UVR photos of aquatic insects of the EPT group in print but substantive proof of their UVR markings. Pinned specimens were used for the photos.

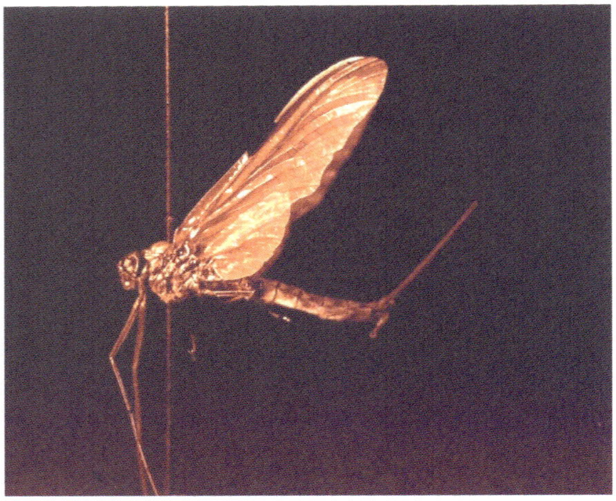

Siphlonurus occidentalis

To the right are three male imagoes (spinners) of the genus *Siphlonurus*, commonly known as Grey Drakes. In visible light the wings of these insects would appear nearly perfectly transparent with no color to remark upon. However, a comparison of the wings in the ultraviolet shows a clear distinction between species. The *S. occidentalis* has high UVR on the forewings overlaid with clean vertical strokes of lower UVR; the hindwings are not patterned. The *S. rapidus* has slightly lower UVR as a base in the wings, with more dark bands on the forewings and dark bands on the hindwings.

Siphlonurus rapidus

The abdomen of *S. occidentalis* is higher UVR underneath than it is above while S. rapidus has light specks of high UVR on a body of medium UVR. The head and thorax of *S. rapidus* are also significantly lower UVR than those of *S. occidentalis*.

Here are the wings shown in greater detail.

Siphlonurus occidentalis

Siphlonurus rapidus

A third member of the *Siphlonurus* genus, *S. typicus*, is also easily distinguished through UVR. Note the clear darker markings on the abdomen. The wing detail shows discrete vertical bands on both the top of the far forewing and the underside of the near forewing with a lighter, unmarked hindwing.

Siphlonurus typicus

Siphlonurus typicus

Below is the *Ephemera guttulata*—often known to anglers as the Green Drake.

Ephemera guttulata

Ephemera guttulata

The body of the *Ephemera varia*, below, is higher in UVR than *E. guttulata* and exhibits bright points of UVR.

Ephemera varia

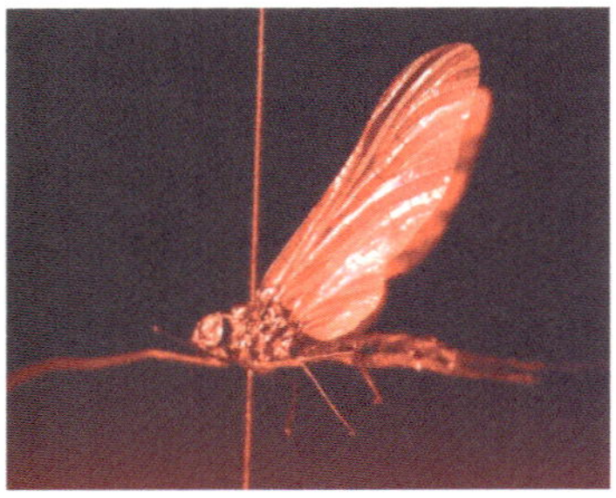

Leptophlebia sp.

It is outside the scope of this book to present all of the mayfly genera in the UV. It is, I think, sufficient to indicate that UVR markings are present on mayflies, that they vary greatly from species to species and between genders, that they have a role in the life and mating of the insects, and that the trout will incorporate the UVR patterns as part of their pattern/template of the prevailing hatch.

Most of the mayflies I photographed had wings with an average base UVR of 30%. Some were as low as 15% while others reached 40%+. Clearly, the acute UV polarized vision of both the mayflies and trout would have no difficulty in distinguishing these markings.

Trichoptera – Caddis

The caddis live a wild and wonderful life compared to the mayflies. A mature caddis may be looking for love for just one week or as much as two months depending upon the species… before expiring. Caddis typically rest in trees and vegetation by night and seek a mate by day. Thus, the UV markings that are necessary to the mayflies that swarm in low-light would be dangerous for the caddis that moves by day, when avian predators with acute UV vision are flying above.

Accordingly, you will note that in the eight species shown below the visible UVR wing markings are suitable for camouflage when the wings are folded. Perhaps entirely different markings are seen in flight or the underside of the wings. The body in most cases appears higher in UVR than the wings.

Psychoglypha subborealis

Rhyacophila vocala

Rhyacophila brunnea

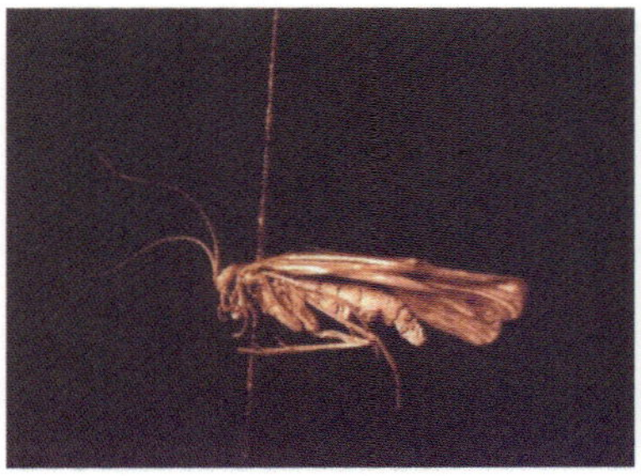

Rhyacophila amicis

Rhyacophila appalachia

Rhyacophila vocala

Pycnopsyche luculentus

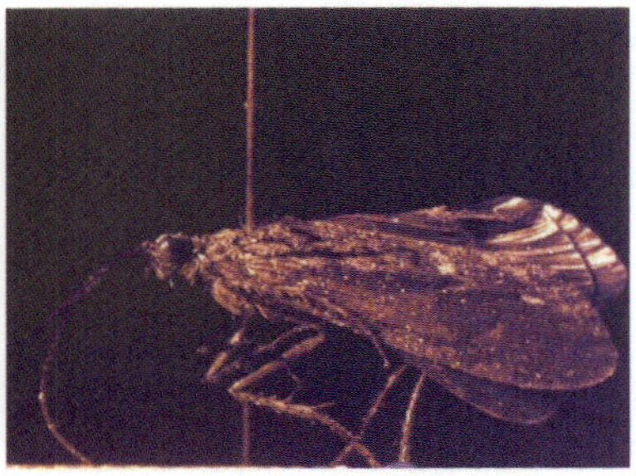

Agrypnia vestita

Lepidoptera

Moths and butterflies often exhibit UVR in their wing markings. In some cases it is the females only, as in the *Aporia crataegi* butterflies below. Note that some reflect well over 50% UV. This UVR carries with it a strong risk of predation by birds, which have been found to select their butterfly prey by UVR clues.[29] Though fly fishermen don't often use imitations of butterflies, any such imitations should include strong UVR on the forewings.

(Figure 1) The presence or absence of UV patterns was determined photographically comparing the wing patterns against a grey scale that was included in all photographs. The amount of UV light reflected (i.e., reflection in %) is indicated by numbers next to the grey scale. This UV photograph shows the upper side of *Aporia crataegi* (four males on the left and four females on the right), whose males appear to absorb UV light (i.e., they appear darker than 12% on the grey scale) while the forewings, but not the hindwings, of females reflect UV light.[29]

Moths

Moths are a favorite large calorie source of summertime trout in the early evening. Nocturnal moths are three times more likely to have UVR markings than daylight feeding (diurnal) species. This makes sense as the nocturnal moths must find mates in low-light, predominantly UV, conditions. The angler therefore should tie his moth imitations with spots of strong UVR in mind. The White Miller is a traditional moth imitation which, quite handily, is tied with highly UVR materials.

Bees

I haven't been able to record the UVR of bees. However, bumblebees do have UVR adornment. It has been recorded that the UVR of bumblebees varies by altitude, with the higher altitude bumblebees reflecting more strongly than those from the lowlands.[30]

Grasshoppers

Grasshoppers *(Orthoptera)* are a staple trout food during the grasshopper season. I have been unable to find studies on the UVR of grasshoppers. We do know, however, that they use color as camouflage. Logically, since the major predator of grasshoppers is songbirds, which have UV vision, we can assume that grasshoppers have the same UVR as their most common environment—grasses. Otherwise, the VIS camouflage is wasted, not a common evolutionary trait. Measurements I've taken in New Hampshire indicate that UVR of green grass, as well as dead brown grass, varies from UV absorbent to a UVR of approximately 30%. This is consistent with the variation I have seen in other foliage. Therefore in reproducing grasshoppers we can use materials that cover that range. I favor taking the high end over the low end.

Above right is one grasshopper who offered to pose for me (no grasshoppers were harmed in the making of this book).

You will notice that the grasshopper has low to medium UVR on its belly and flank, but the wings, leg, and head are UV absorbent. This makes sense; the major predator species would see the 'hopper from above—so it has low UVR from that angle—whereas another grasshopper would see it at the same level and be able to interpret the markings.

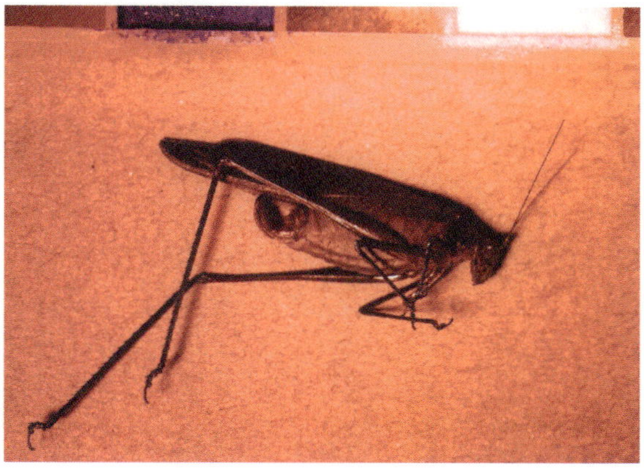

Ants

Ants *(Formicidae)* seem to be a choice delicacy for most trout. Ants were also the first animals known to use ultraviolet vision—Sir John Lubbock wrote of this in 1882. I have been able to take photographs of only one species of black ants—of some eight thousand ant species. While these black ants had no UV markings, we cannot assume that this would be universally true. Just as ants come in many colors—e.g., red, red and black, honey, cinnamon, amber—it is likely that they vary in UVR as well, given their differing environments.

Beetles and Crickets

Crickets and beetles, are the other leaders in terrestrial insects common as trout food. Among beetles, UVR is often found in areas of pubescence (areas covered with fine short hairs).[30] To err on the side of caution, I would suggest little if any adornment of ant, beetle, or cricket imitations with UVR reflective material.

Mosquitoes

Don't forget mosquitoes—you may wish to, but they seem to make their presence known on stream and pond just as the fishing is at its peak. Mosquitoes are sensitive to UV, hence the UV lights used in bug traps and zappers.[31] The silvery-white bands on mosquitoes strongly reflect UV[30] and we should mimic this in our flies—which the old Mosquito dry fly pattern did very well.

Damselflies and Dragonflies

These insects are commonly seen hovering over trout waters, snatching up smaller insects.

(Figure 19) UV photograph of a mature male (left) and a mature female (right) *Plathemis lydia* and mature male *Libellula pulchella* (below) (*Odonata: Libellulidae*), illustrating sexual dimorphism and interspecific differences commonly found among dragonflies. The areas that reflect UV also reflect "visible" light, but not as intensely; the reflection is a "Tyndall effect."[30]

UVR is found on the iridescent body regions of some male and female damselflies as shown in Figure 19 above. Strong UVR is present on *Libellulidae*, the largest family of dragonflies, especially on the abdomens of mature males.[30] The bottom insect in Figure 19 exhibits powerful UVR markings, seen as the white on wings and abdomen, and equally bold patches of UV absorbance, as evidenced by the black on the wings. Since the major predators of these insects are birds with UV vision, it is notable that the male dragonflies have significantly more UVR markings than the females. As with many other species, the males are expendable (sigh). Another illustration of the pronounced UVR of some male *Libellulidae* is Figure 2.[32]

(Figure 2) Three species of *Libellulids* illuminated under 300-400 nm U.V. light. Percent reflection indicated in lower left corner: from left to right, 1%, 2.5%, 5.0%, 16.0%, > 25.0%.

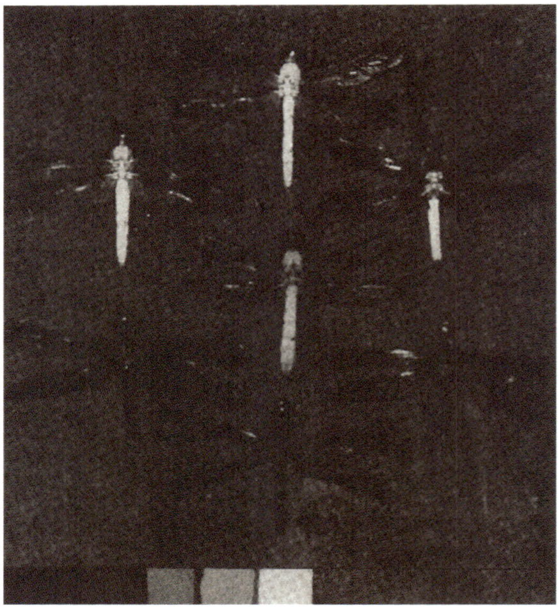

CHAPTER TEN

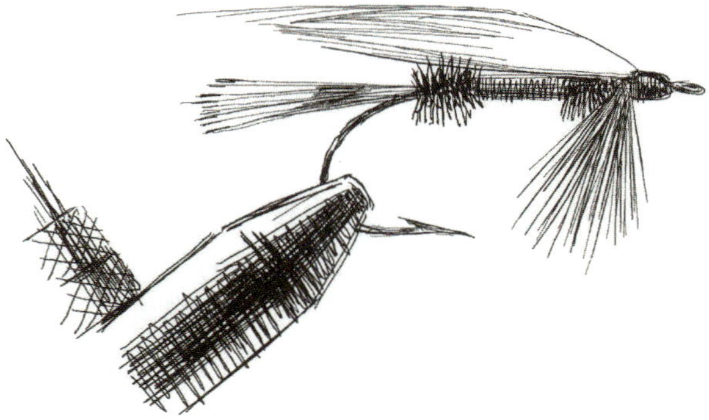

ARTIFICIAL FLIES

Common sense tells us that the most popular artificial fly patterns should be those that consistently catch fish. Of course, that is not always the case; fishermen follow foolish fads as often as any other man or woman. However, assuming for the moment that anglers are made of better stuff: if UVR is really an important factor in getting strikes from trout, the most popular flies will exhibit UVR, where appropriate, or have other characteristics which evoke anger and/or curiosity strikes.

In order to get an unbiased representation of artificial flies to test under UV, I used the suggested dry fly assortment from the *LL Bean Fly-Fishing Handbook* by Dave Whitlock (Winchester Press, Piscataway, NJ, 1984). Many of these fly patterns date from the 1930s or earlier; they have stood the test of time.

Pattern lists from any later than the mid-1980s would usually not include variants, bivisibles, or spiders—all very effective styles. The late 1980s was also known for the beginning of the shift to synthetics and away from natural furs and feathers, introducing the variables of differing material compounds, manufacturing techniques, dye lots, etc. Following are the 22 standard dry fly patterns for trout recommended by Dave Whitlock. Each fly is photographed first in VIS with a neutral background and then in UV without a background. By removing the background in UV there is no reflection but that from the fly itself, making it easier to identify the true UVR.

While the yellow pigments in the lens of the human eye, and the eyes of many other creatures, absorb near-ultraviolet light, some yellow lens pigments do not absorb in the wavelengths critical to UV vision.[41]

#1 – Adams

 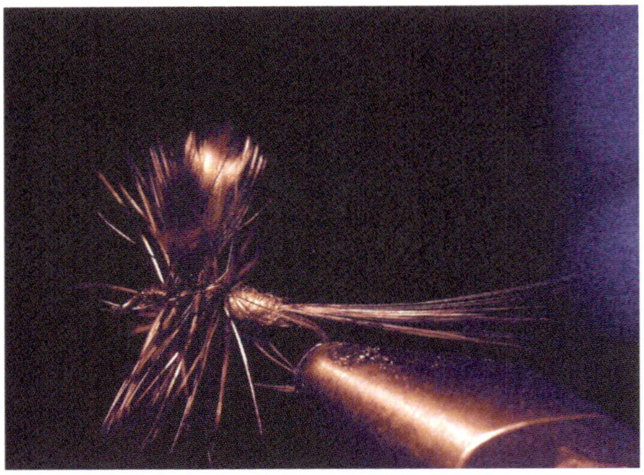

Wings: Grizzly hen hackle tips
Tail: Brown and grizzly hackle
Body: Muskrat fur
Hackle: Brown and Grizzly
UVR: Grizzly hackle, wings, body, and tail. Superfine—dyed natural hair—dubbing was substituted for muskrat fur for the body. This is a common substitution. With muskrat the body would have very low UVR.

The Adams is almost a universal trout fly. It works as well in New Zealand and the UK as in the US. Whether you think it represents a specific fly, or a generic mayfly, the trout are generous enough to take it even during a hatch.

You will note that the Adams as shown is not the original down-wing caddis imitation. Few use that tie anymore, preferring the upright wing. The full hackle, and use of grizzly for wings, hackle, and tail, provides a significant UVR signature without being flashy. Just a little bit in each section. Note this approach when designing generic patterns.

#2 – Light Cahill

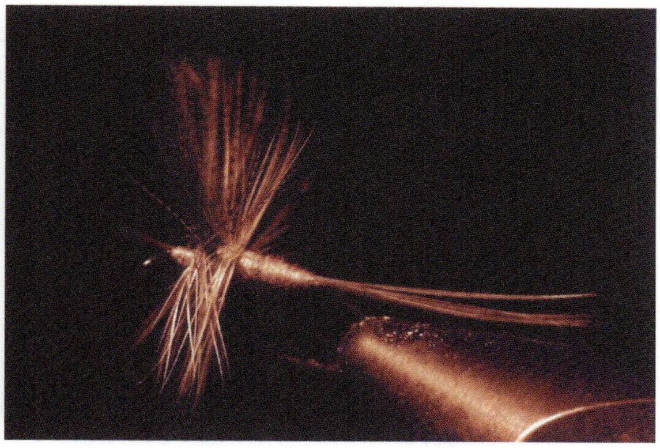

Thread: Pale yellow
Tail: Light ginger to cream hackle
Body: Light ginger or cream fox dubbing
Hackle: Light ginger to cream
Wings: Wood duck flank
UVR: Hackle, body, and tail.

The Light Cahill is a favorite of many fishermen. There are enough hatches of light-bodied mayflies to warrant its inclusion in almost anyone's fly box. Like the Adams, the pattern as tied above has substituted a cream-dyed Superfine fur dubbing for the fox. In this case the difference in UVR between the two materials would be minimal.

The color of thread is important in the design. Here it allows not only a light body—a dark thread would show through when wet—but provides a head with high UVR. One of the drawbacks of the commonly accepted "Catskill style" is the apparent fixation with a diminutive head on dry flies. Here is a case where seeing the head might be an advantage. The small head doesn't show well in VIS above, but it is obvious in UVR. I would make it even bigger, not to the size of a Stimulator, but definitely bigger.

The Light Cahill was originally designed by Dan Cahill in 1880 as a wet fly to imitate the *Stenonema canadensis* mayfly. The present dry fly dressing is traced to Wm. Chandler of Neversink, NY. Anglers use this fly to imitate almost any light-bodied mayfly—from the large Yellow Drake to the smaller Pale Morning Duns, Pale Watery Duns, and Pale Evening Duns. Art Flick said of the Light Cahill "… it is probable that more fish have been taken on this than on any other pattern of dry fly." (*New Streamside Guide to Naturals and Their Imitations*, Crown, NY, 1969). As an early dusk fly this is especially effective as most of the mayflies it imitates hatch in late afternoon and into the night.

Quite a powerful UVR statement is made by this fly. Naturally, with the strength of its UVR, the Light Cahill declares itself in even the lowest VIS conditions, when the UV will dominate.

#3 – Hendrickson

Thread: Light
Tail: Natural blue dun hackle (this tie has microfibbets)
Body: Pink tinged fox fur (urine stained)
Hackle: Natural blue dun
Wings: Wood duck flank
UVR: Hackle has medium UVR; head, body and tail have strong UVR.

The Hendrickson, here with the expected fox fur dubbing, has many of the virtues of the Light Cahill but is subtler. Art Flick (ibid.) considered the Hendrickson as an excellent imitation of the female *Ephemerella subvaria* and used the Red Quill, identical except in body, as the male *E. subvaria.*

The Hendrickson is our first, but not last, instance of the notorious "Blue Dun Conundrum." Seldom can any two people agree on the color known as natural blue dun. Mr. Art Flick himself states that both the Hendrickson and the Red Quill use the same natural blue dun hackle, but the photos of the flies he tied for his seminal work have completely different color hackle. It is not a matter just of saturation, the hue itself is different—one is clearly a medium brown-gray, the other is certainly a light blue-grey. For our purposes, however, we will identify the natural blue dun as follows:

A Light Blue Dun from Metz looks liks this in VIS:

and this in UVR:

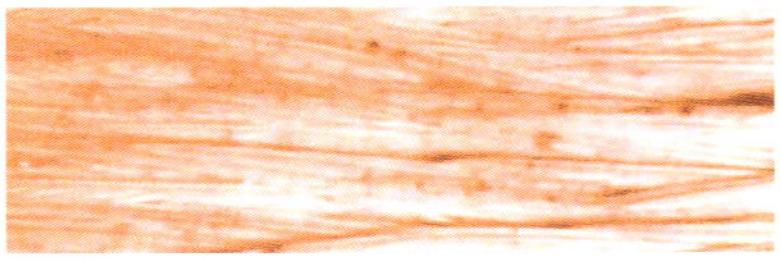

A Medium Natural Blue Dun from Metz in VIS:

and in UVR:

Examining the dictionary definition of "dun" gives us a starting point. "of a horse: having a grayish-yellow coat with black mane and tail." (Merriam-Webster Online)

The use of the term "blue dun" in fly tying predates the common use of the automobile. Needing a color description, it seems that the tiers turned to what would be most common—horses. The tiers used acknowledged horse colors as reference points. Thus, when documenting a fly pattern, "dun" would be broadly understood as being a "grayish-yellow." If in doubt, one only had to look out the window at horses going by in the street to understand the tier's intent.

Below are six color patches taken from photos of dun horses. The color designations are those provided by the horse breed standards.

Med. Blue Dun

Blue Dun

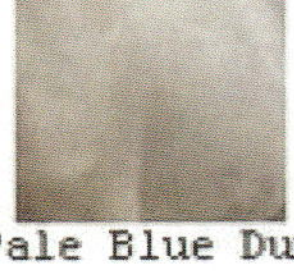
Pale Blue Dun

Dun

Red Dun

Light Dun

#4 – Black Gnat

Thread: Black.
Wing: Slate gray quill.
Tail: Black cock hackle fibers.
Body: Black chenille.
Hackle: Black cock hackle.
UVR: Wings.

The Black Gnat in small sizes is used to imitate midges, in medium sizes, the biting midges, and in larger sizes, perhaps a horse fly. In New Zealand the same fly with the same name is used to imitate blow flies. In all cases, the insect imitated is two-winged and black or almost black in body.

As you see in the images above, the UVR is concentrated in the wings and is medium in intensity. The Black Gnat is the only fly in this list with fragile upright quill wings. So, if fly patterns evolve over time through testing different combinations, the fact that the Black Gnat still has these wings is significant.

#5 – Blue Wing Olive

Wing: Medium Blue dun hackle tips.
Tail: Medium Blue dun.
Body: Olive fur.
Hackle: Medium Blue dun.
UVR: Hackle, wings, body, and tail.

One thing that may be noted about the Blue-Winged Olive is that so many tiers omit the wings entirely. Perhaps the wings don't add much to the VIS, since they are the same material as the hackle, but they certainly reinforce the UVR of the hackle. With some fly patterns the hackle is far more apparent in the ultraviolet than the wings and you might be tempted to not add wings. While this might work under low-light conditions, we would be foolish to regard the UVR as the determinative factor when winging a fly for daytime use where all the colors come into play.

#6 – Quill Gordon

Thread: White
Wings: Wood duck flank, tied upright and divided
Tail: Light blue dun hackle fibers
Body: Peacock quill taken from eye feather reinforced with fine gold wire
Hackle: Light blue dun hackle
UVR: Hackle, white of body, and tail.

The Quill Gordon is another case of being Dunned—though not in the financial sense. Preston Jennings in his superb work *A Book of Trout Flies* (Crown Publishers, NY, 1935) calls for light blue dun hackles and tail. Art Flick presents the pattern in *A Streamside Guide* with medium blue dun. Of course, that is not the end of the matter: Ray Bergman in *Trout* asks for "bronze blue gray," Dave Hughes in *American Fly Tying Manual* wants "rusty blue dun," John Merwin writes in *John Merwin's Fly-tying Guide* of simple "blue dun", and so on, *ad infinitum.*

Notice the interesting UVR of the Quill Gordon. The body glows with contrasting stripes. As evening draws on, this fly would come into its own. Yet, the Quill Gordon is effective even when and where no hatches of the mayfly occur. Why? Perhaps because with its "silvery-white bands" reflecting in the UV, the Quill Gordon makes an excellent mosquito imitation! Try it next time the "mossies" are biting and the trout are not.

#7 – March Brown

Thread: Orange
Wing: Dark Heavy Barred Wood Duck Flank, Upright and Divided
Tail: Reddish brown cock hackle fibers
Body: Red fox underbelly fur
Hackle: Reddish brown and gray grizzly cock hackles mixed
UVR: Head, grizzly hackle, and body.

If you ever wondered what a difference thread color makes, the March Brown is the exemplar. The orange thread helps with the body when wet and is a great accent to the head. Note how the fox fur body, picked out a bit, appears much more substantial—and succulent—in UVR. The grizzly hackle, as with the Adams above, gives us plenty of UVR highlights.

#8 – Gray Fox Variant

Thread: Primrose (pale yellow)
Wing: None
Tail: Ginger cock hackle fibers
Body: Stripped ginger cock hackle quill
Hackle: Light gray grizzly, light ginger, and medium ginger cock hackle, tied over each other and bunched close, 2 sizes larger than hook size
UVR: Medium UVR from quill body, ginger hackle and tail, strong UVR from Grizzly Hackle.

I just checked the 2009 catalogue of a company that purports to be the "largest producer of hand-tied flies in the world." The company sells no traditional variants, bivisibles, or spiders.

Nevertheless, the traditional variant is a very effective fly style. Essentially, a variant is a traditional upright hackle fly with much longer, 1.5-2.0 times, hackle and tail. If the tail is too short, the fly will not cock properly. Conversely, if the tail is too long, the fly will tip onto its face. The original variants had tiny upright wings, half the length of the hackle. Soon, however, the wings were omitted as superfluous as they were scarcely visible through the depth of hackle.

"Hand-tied." How else are flies tied? All artificial flies which are tied are tied by hand. There would be no practical ROI in building machines to tie flies when superb flies are produced for pennies in distant parts of the world.

#9 – Brown Spider

Thread: Brown
Wing: None
Tail: Brown
Body: Stripped brown hackle quill
Hackle: Brown, 2-3 sizes larger than usual
UVR: Low UVR from the quill body; virtually none from hackle, and tail.

It is very difficult to find good hackles for spiders. The best hackle is spade hackles from the rooster's throat. These are seldom available.

> "Now genetic dry-fly hackle from various companies is about all you can find in most fly shops. For the most part, the quality is better now than it's ever been, but the one flaw is that the spade hackles have been all but bred out of these necks.
>
> Spade hackles are the short, wide throat feathers that were once found at the edges of the widest part of a skinned cape. They're too stubby and wide to be used as wound hackle on anything but spider patterns, but the barbs are long, shiny, and stiff. They're the best, if not the *only* material for tails on many standard dry fly patterns.
>
> ... for quite a while I was reduced to mining old picked-over necks for spade hackle and buying second- and third-rate so-called dry-fly necks that were just about useless except for some good tailing feathers."

– John Geirach from *Good Flies* (Lyons Press, NY, NY, 2007)

Some report good results in using Coq de Leon rooster feathers for spiders. I have no experience with Coq de Leon feathers; I still stumble along using spade hackles from old capes and saddles. For tails I most often use the guard hairs from bleached beaver pelts.

"Of all trout flies, the one you can least afford to be without is the spider. It can be a Brown Spider, a Badger Spider, a Blue Dun Spider, or whatever kind you like, but at least one of these high riding floaters should be in the kit of every serious fly fisherman. The spider is thought to imitate long-legged insects like the crane fly, spider, or water strider. True or not, they seem to suggest insect life in general because they are usually effective even when caddis flies, midges, and mayflies are on the water. In appearance, the spider-type dry fly consists of a relatively small hook tied with extra large hackle. Although they are sometimes made without bodies or tails, in my experience a fly so tied is much less effective. When the spider is tied with a tiny set of wings it is properly called a "variant," but in recent years the terms have become arbitrary, and many variant patterns are now wingless. The great attraction of this fly is its ethereal, almost lifelike ability to dance across the surface. It will even turn around on the water, and the slightest change in current or puff of wind will send it skating for a few inches before floating again. The secret of course, is in the hackle, which may be two inches or more in diameter, depending on the hook size. Hook and hackle are out of proportion to each other, the hackle being roughly three or more times greater than the hook would regularly require in an ordinary dry fly.

The reactions of trout to a spider-type fly, are different from their responses toward ordinary flies. Logical behaviour among trout is nothing more than a pleasant myth at best; the spider, for instance, commonly causes the trout to rise above the surface and take the fly going down. Now whether this is because the fly actually excites them, or because the sensitive movement of a spider simulates an active insect about to take off, I am not sure. I have caught many trout this way, and over a long period of time the only impression left me is that the fish were anxious that the fly should not escape."

– A.J. McClane from *The Practical Fly Fisherman*

#10 – Dun Variant

Thread: Olive
Wing: None
Tail: Medium blue dun cock hackle fibers, long
Body: Reddish brown cock hackle quill
Hackle: Medium blue dun cock hackle, 2 sizes larger that hook size
UVR: Very strong from the hackle and tail.

You can see the hackle much better in the UVR image. Imagine that floating down gently onto still water, or bouncing over riffles—irresistibly enticing!

#11 – Brown Bivisible

Thread: Brown
Wing: None
Tail: Brown cock hackle tips tied in as tails prior to palmering
Body: None
Hackle: 2 Brown cock hackles palmered for 2/3rds shank length, then faced with white cock hackle
UVR: Very strong from the white hackle face.

The Brown Bivisible is purported to have been developed by Edward Hewitt in 1898 for the Catskills, according to Harold H. Smedley (*Fly Patterns and Their Origins*, Westshore Publications, MI, 1946), although Smedley infers a possible Michigan origin.

Whoever was responsible for the pattern, or style, since Bivisibles are tied in every color including pink, it is a great fly. Most anglers liked the pattern for two simple reasons: the profuse palmered hackle kept the fly floating high even in rough water, and the white front made it easy to see. Personally, I think the trout liked it because it gave a strong blaze of UVR. It has fallen out of favor not because it failed to catch fish, but because it does not represent a specific insect.

You wouldn't think that a pattern tied in 1900 would be one of the most difficult flies to tie properly today. Nevertheless, the correct tie of the Bivisibles has the head higher than the tail without the hackle being trimmed. This was possible because dry fly hackles came from gamecocks from India or common chicken breeds. These hackles were short in overall length, had lots of web, and the individual barbs were short near the tip and much longer near the butt. All one had to do was to tie the tips as the tails and palmer two hackles forward; the taper of the fly was natural to the feather. Today we have the superb hackle produced by farms committed to lines of birds raised solely for their dry fly hackle. Most of the hackle has a uniform barb length, a fact we all enjoy—until we try to tie a Bivisible.

#12 – Adams Parachute

Pattern: Same as Adams but tied parachute with Antron post. This has a dyed dark grey hare's ear dubbing body. Slightly lower UVR of the body than that of the upright Adams (#1). UVR: Medium UVR from the grizzly hackle, very strong UVR from Antron post, medium UVR body, and strong UVR tail.

For a long while I tied my parachute flies with a fluorescent green/yellow Antron post, as this color was easier for me to see in fast water. I believed that the trout wouldn't care since the post would be barely visible from below. How wrong I was. That wing glows in the UV! So I've switched back to the white Antron, which more nearly resembles a fly's wing color in the VIS. I may even make my posts taller for both the trout's benefit and mine. Note to self: Why do fish never appreciate all my hard work?

#13 – Irresistible

The Irresistible is a hard pattern to identify as it has become more a style than a pattern. The recipe below is an average of many slightly different recipes.

Wings: Brown Bucktail or Elk.
Tail: Brown Bucktail or Elk.
Body: Gray-brown deer hair from body of whitetail deer, clipped to taper.
Hackle: Medium blue-dun cock hackle.
UVR: Hackle, body, and some tail.

This is a lot of juicy fly, whether in the VIS, the UVR, or both. While there may be better searching flies and better night-time flies, this fly has its moments throughout the day, when the fish want something "buggy." Personally, I would tie it with grizzly hackle for general use. The Irresistible has been replaced by the Humpy in many fly boxes because it is easier to tie and gives the same outline. It also provides much the same UVR signature.

#14 – Hairwing Royal Coachman

Originally the Quack Coachman or simply The Quack. With a meatier body, commonly called the Royal Wulff

Wings: White calftail.
Tail: Brown cock hackle fibers.
Body: Peacock herl divided by a band of red floss.
Hackle: Dark brown cock hackle.
UVR: Wings.

The white calftail wings are highly UVR while the rest of the pattern is very low UVR. The wings are often much fuller than in the tie above, resulting in an even stronger UVR signature. We have already spoken of this fly, but it is such a popular pattern that we will discuss it again later in the book.

#15 – Henryville Caddis

usually known as Henryville Special

Wing: Medium Gray Mallard quill segments, flaring outward.
Underwing: Wood Duck flank fibers.
Tail: None.
Body: Olive Floss with gray grizzly hackle palmered.
Hackle: Brown cock hackle.
UVR: Wing has strong UVR, underwing has strong UVR, brown hackle has medium UVR, and grizzly hackle has high-medium UVR.

This fly is believed to have originated in Henryville, PA, in the 1920s. Many pattern books give the hackle as ginger, which I prefer. The down-wing of this fly shouts "Caddis" and it is one of the most popular of the caddis patterns in New England. The quill wing is fragile, however, so hair-wing caddis patterns have supplanted this fly for certain waters, the Elk-Hair Caddis being the best known. The Elk-Hair Caddis is excellent for fast water, but in glassy water the more clearly defined shape of the Henryville seems to draw more fish. Personally, for fast water I prefer the Wright's Royal which floats well, is easily seen by these old eyes, and has the classic "Am I a flying ant or a caddis?" signature—oh, and the Royal Coachman UVR as well.

#16 – Black Ant

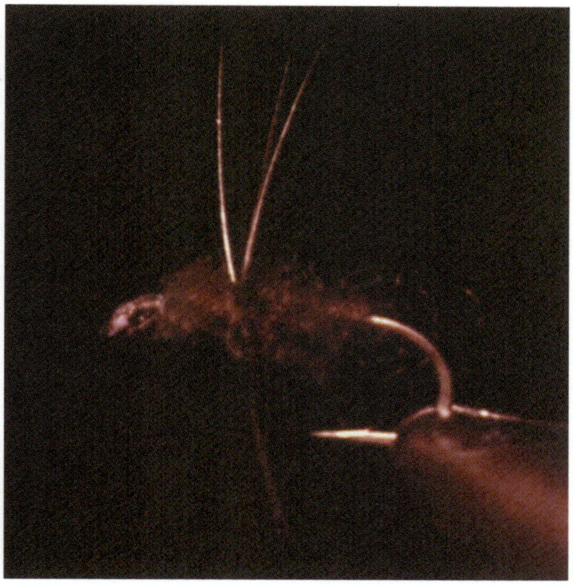

Thread: Black.
Body: Black fur dubbing, one larger behind hackle, another in front of hackle.
Hackle: Black, upright, somewhat forward of center of hook.
UVR: Very little UVR.

This black ant pattern doesn't float as well as a parachute or a McMurray ant, but it doesn't have the profusion of legs of the first or the hard outline of the second. Low/no UVR is also appropriate for the insect.

#17 – Cinnamon Ant

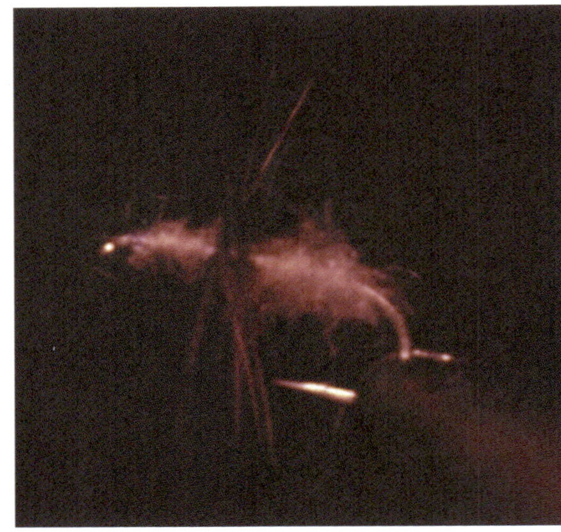

Thread: Medium red-brown.
Body: Cinnamon fur dubbing, one larger behind hackle, another in front of hackle.
Hackle: Cinnamon, upright, somewhat forward of center of hook.
UVR: Strong UVR in body. I have been unable to photograph any cinnamon ants in UV, so I am uncertain whether they have UVR markings.

#18 – Beetle

Head: Black deer body hair spun and clipped.
Collar: Black deer body hair spun.
Body: Black dubbing.
Wing: Black quill.
UVR: Some specular reflection from collar, but essentially no diffuse reflection.

This is a rather elaborate beetle pattern which appears to be a smaller, shorter Letort Cricket.

#19 – Dave's Hopper

Tail: Red Bucktail
Wing: Mottled Turkey Wing
Underwing: Pale Yellow dyed Deer Hair
Body: Yellow Antron or Poly Yarn, or wool
Ribbing: Brown or Cree Hackle, Palmered
Head: Natural Deer Hair, Spun and clipped
Collar: Natural Deer Hair, Spun
Legs: Cock Pheasant Tail Quill Barbs
UVR: Head, collar, and wing.

This is a popular interpretation of the Dave's Hopper. Dave Whitlock originated the pattern but it has become so popular that numerous variations now carry the name—some designed by Dave himself. The original, or at least the pattern tied by Dave Whitlock in *Art Flick's Master Fly-tying Guide* (Crown, NY, NY, 1972) has the spun deer-hair head trimmed like a bass popper with the top of the head projecting more than the bottom. Otherwise, the materials are as shown above. I usually omit the fancy legs.

With medium UVR from the head and wing this fly is replicating a subset of grasshoppers living in green grasses. Yet this pattern catches trout even when grasshoppers are not present and in areas where the grasses have much lower UVR. The second most popular grasshopper imitation, Joe's Hopper, also has a UVR head (grizzly hackle mixed with brown) and medium UVR wing (mottled turkey quill) and, as above, a low UVR yellow body. The two patterns do not look similar in VIS, but do in UVR. Curiouser and curiouser…

#20 – Jassid

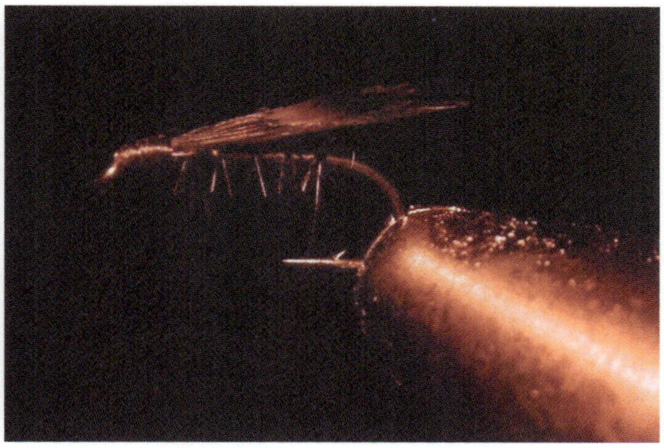

Body: Black hackle, palmered and clipped on top
Wing: Jungle cock eye.
Head: Black thread.
UVR: Essentially none.

This is one of many flies with which I have no experience. It grew from the work of Vince Marinaro who fished the limestone streams of Pennsylvania. I do not know what the insect it approximates has for UVR markings, if any.

#21 – Green Inchworm

 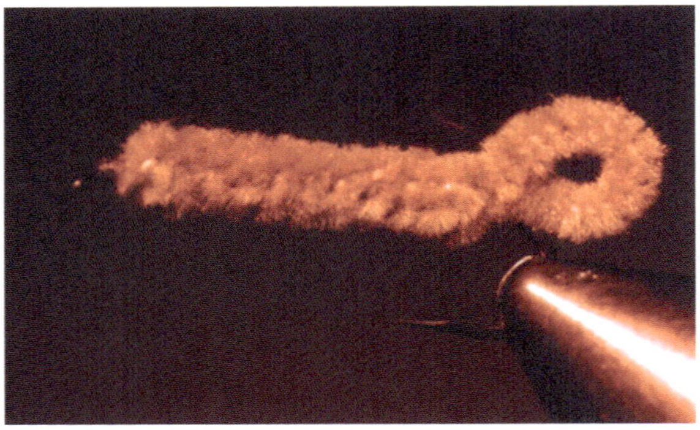

Body: Chartreuse chenille.
Tail: Chartreuse chenille.
UVR: Body and Tail, including spots where the bright core thread shows through.

This fly (fly?) is just plain fun. Inchworms we all know from childhood, those short green caterpillars that stroll along your arm in a disarming manner. I don't know whether trout take this fly as the caterpillar, or just because it is so bright in the UV. Nevertheless, take it they do.

#22 – Cricket

 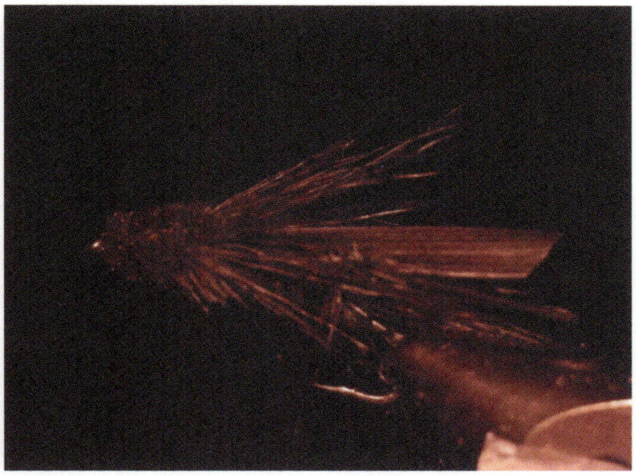

Body: Black fur dubbing.
Wing: Black goose quill.
Head: Spun black deer body hair, clipped.
Collar: Spun black deer body hair.
UVR: None.

This is the Letort Cricket, a larger, longer version of the beetle seen previously. Again no UVR; which seems an apt representation of the average garden-variety cricket.

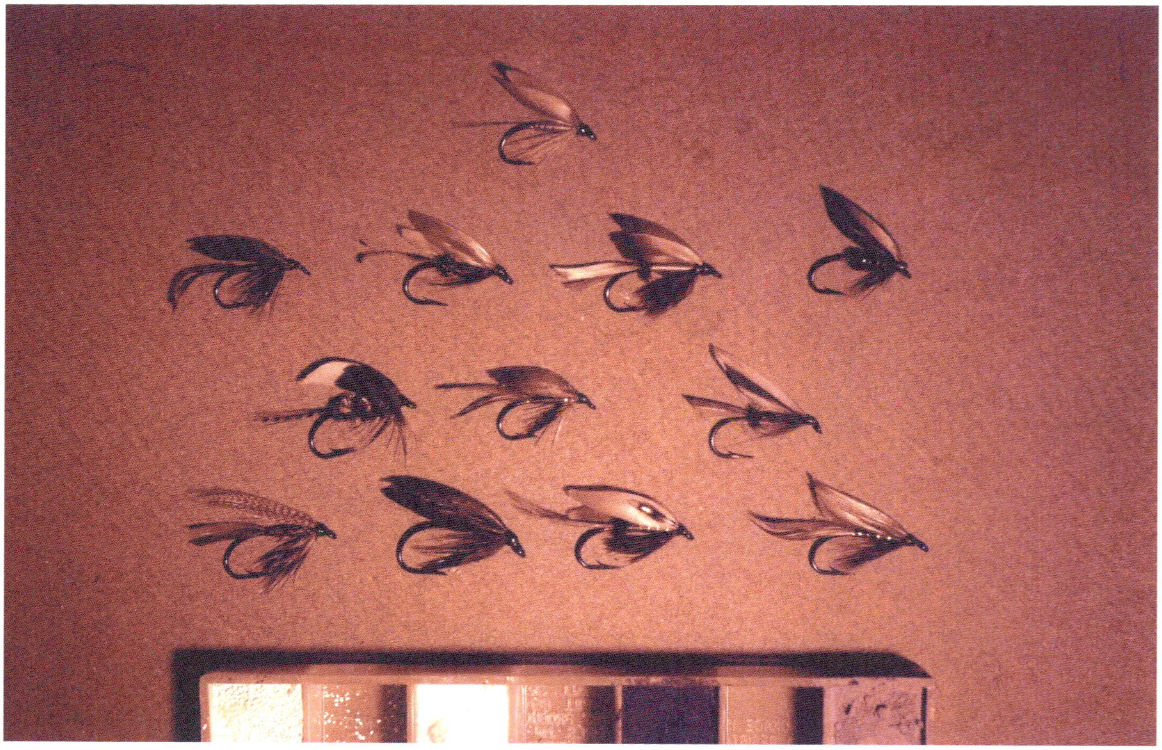

You may assume after reading this chapter that all imitative artificial flies have some UVR, apart from the obvious beetles and crickets. That is not the case; many imitative mayfly patterns have been tied with no UVR; however, they have passed into oblivion. Who fishes the Black Angel—an all-black fly similar to a Black Gnat but with black wings instead of the UVR gray wings? It had its moment, but was found wanting. What about a Yellow Sally dry fly? I have never seen the pattern, but the Yellow Sally is useful as a wet fly, so I assume someone tried it as a dry. But there is the difference: many of the brightly colored but low UVR wet flies, such as the Red Ibis and Yellow Sally, are great lures. However, when fishing a drag-free drift in imitation of an insect, the pure, non-representational lure is not as effective as an insect imitation.

To the left is a group of traditional, colorful wet flies in the VIS, and in the UVR. Note that the white-tipped mallard wing of the McGinty has the highest UVR. The natural white duck quills of the Royal Coachman and the White Miller (top fly) have lower UVR than the natural white goose quill sections of the Parmachenee Belle. While all of the above flies work subsurface, only three of them are commonly used as dry flies—the White Miller, the Royal Coachman, and the McGinty. All three are imitative. The White Miller is an evening moth imitation, the McGinty a bee, and the Royal Coachman... we are working on that one.

Some flies which are only imitative in an abstract or representational sense are nevertheless effective in taking trout when fished dry. That the Bivisibles are no longer fished is not so much a matter of their kill-rate, which is great, as that they don't represent a particular insect. If I needed a searching pattern for low-light on fast water, a light blue dun bivisible might be my choice.

As I stated at the beginning of this chapter, "...if UVR is really an important factor in getting strikes from trout, those flies will exhibit UVR where appropriate or have unique characteristics which evoke anger and/or curiosity strikes." How well did these 22 flies perform against this test? All but one had the appropriate amount of UVR; that was the Brown Spider. However, as A.J. McClane noted above, "The reactions of trout to a spider-type fly, are different from their responses toward ordinary flies." If the strikes on spiders are from curiosity or "anger," rather than careful feeding on an imitation of an insect then all the flies met the test criteria.

Today we tie most of our dry Mayfly imitations on standard length hooks, but it was not always thus. In the early days of dry fly fishing in the U.S., the bodies of dry flies were tied to the approximate length and bulk of the natural insect. Of course, these larger flies were much more difficult to keep afloat than a thin-bodied fly on a much shorter hook.

Try tying some dry patterns with bulkier bodies on 2x long dry fly hooks and compare your angling success rate against the more delicate tie.

The images below are from *The Natural Trout Fly and Its Imitation* by Leonard West (1912).

PLATE 4.

EPHEMERA (DUNS—SPINNERS), ETC.

No.	Name.	Eyes.	Legs.	Thorax.	Wings.	Body.	Rib.	Tails.	Notes.	Hook.
36	March brown	——	Partridge & red cock	Brown herl	Hen pheasant or bittern	Brown quill	Gold tinsel	Cock pheasant	Bittern, partridge or quail wings do well at times.	CG
37	Early olive dun	Red wool	Olive cock (dyed)	Olive quill	Starling (dyed)	Olive quill	,,	,,		BD
38	Small red spinner	Yellow	Prismatic	Sienna quill	Pale starling	Burnt sienna quill	,,	,,	(Red, yellow, and green hackle	BD
39	Red spinner	Brown wool	,,	,,	,, or blue hackle points	Burnt sienna quill	,,	,,		CF
40	Yellow tail spinner	,,	,,	Dark brown	Pale starling	Yellow and dark brown quill	,,	Golden pheasant		BD
41	Green dun	Green wool	Green cock	Dark green	Medium starling	Green quill	Gold wire	Green cock		AB
42	Green spinner	,,	Prismatic	Olive green	——	Dark olive	,,	Golden pheasant		AB
43	Olive dun	,,	Olive cock	Olive quill	Pale starling	Olive quill.	,,	Olive cock		BC
44	Olive spinner	Dark olive	Prismatic	,,	,,	,,	,,	,,		BC
45	Jenny spinner	Red wool	,,	Black ostrich	——	White horse-hair	——	Blue cock	A touch of red wool at tail	AB
46	Medium olive dun	Green wool	Olive cock	Olive quill	Olive starling	Olive quill	Gold wire	Olive cock		BC
47	Pale evening dun	Dark green	Blue cock	Ostrich	Medium starling	White horse-hair	——	Blue cock		AB
48	Dark olive	Green wool	Dark olive	Dark olive	Dark olive starling	Olive quill	——	Olive cock		BC
49	Whirling blue dun	Red wool	Palest olive	Pale olive	Medium starling	Pale olive	Gold tinsel	Cock pheasant		DF

CHAPTER ELEVEN

NATURAL FLY TYING MATERIALS

Feathers

Feathers are the mainstay of the fly tier. The hackles of a dry fly, the wings and tails of most wet, dry, and streamer flies—all of these use the feathers of birds gathered from dozens of countries and climes. The colors are sometimes subtle, sometimes dazzling. But what I do not see when I run my hands over a Silver Pheasant skin is the ultraviolet light reflected from the feathers—and those reflected wavelengths may be crucial to the effectiveness of the fly I tie.

Scientists have studied every family of birds for UV reflective plumage. They concluded:

> "Our results demonstrated that reflectance of incident UV radiation varied markedly across the range of plumage patches sampled in this study: 63.3% of sampled feather patches reflected a significant proportion (more than 5%)of UV light, with 34.6% reflecting greater than 10% and 13.6% reflecting in excess of 20% of incident UV wavelengths. We confirmed that no visual colour class represented in bird plumages was always associated with UV reflectance, just as no visual colour class always lacked UV reflectance (Burkhardt 1989).
>
> Nonetheless, some generalizations emerged (figure 1). The majority (more than 50%) of black, brown and red feathers that we sampled reflected less than 5% of the UV wavelengths. By contrast, 50–75% of yellow and green patches examined had greater than 5% UV, and more than 75% of blue patches had a UV reflectance of higher than 5%. Nearly all white featherpatches reflected UV wavelengths, with very few (2%) reflecting less than 5% of UV light, and a high proportion of both white and blue feather patches reflected greater than 10% UV (690 out of 781 and 55 out of 88, respectively)."[14]

While trout can certainly discern small particles with minimal illumination, a UVR of 5% would be, for the angler's purposes, non-reflective. Throughout this book we will use the criteria employed by the Lepidopterists—15% or greater UVA reflection is treated as UVR, less than 15% is considered UV absorption. The work cited above helps us understand that no natural feather can be determined to reflect UV based solely on its color. While many yellow pigments absorb UV, the UVR of natural feathers is carried in the structure of the feathers, not the pigments. So a natural yellow feather may have a high UVR while a white feather from the same bird dyed to the same VIS yellow will be UV absorbent.

"The domestic cock, from which the hackle feathers are taken was possibly the object of man's first attempt at husbandry. Just when or how these birds were first tamed is, of course, lost in the maze of generations which have followed since man first domesticated the wild jungle fowl of India, Ceylon, and the Malay Peninsula."

– *A Book of Trout Flies* by Preston Jennings (Crown Publishers, NY, NY, 1934)

Hackles

Let's look at some natural dry fly hackles first. Natural hackles represent an anomaly in the fly-tying world. Most of our natural materials are taken from either feral animals—bear, fox, mink, muskrat, opossum, deer—or domesticated animals as a byproduct—calf, turkey, and duck. However, the finest fly-tying hackle comes from birds raised and dramatically altered through selective breeding, specifically to provide materials for flies.

As we begin examining tying materials for UVR, any suppositions as to which materials will reflect UV well will be swiftly deflated. For example, you might think that the glossy Golden Ginger neck (below) would strongly reflect UV … but it doesn't.

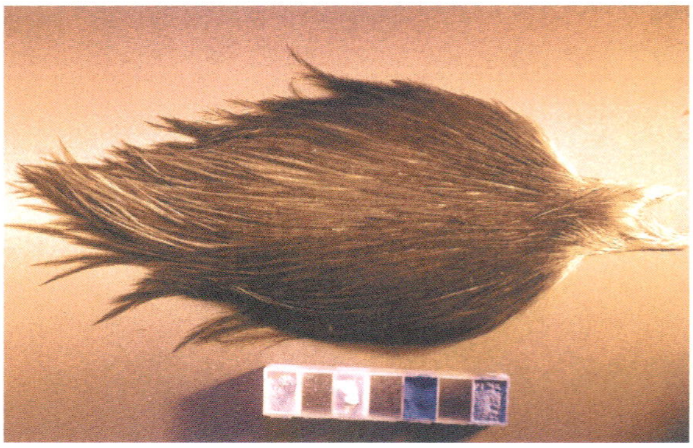

And this Brown neck certainly is dark in the UV.

On the other hand, this Grizzly neck has good UVR from the white fibers—

The Medium Blue Dun below is moderately reflective—

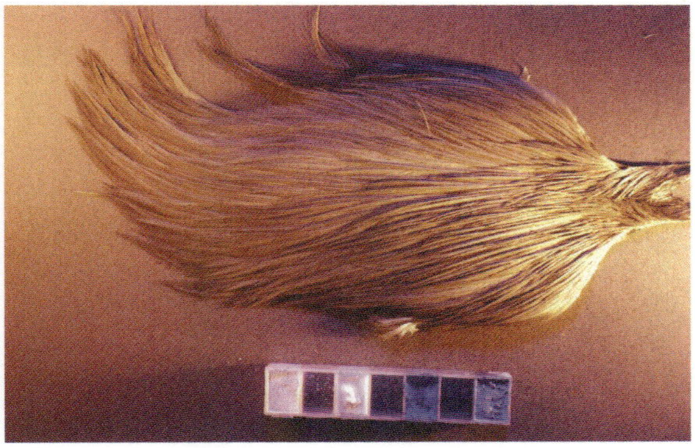

while this Light Blue Dun is highly reflective—

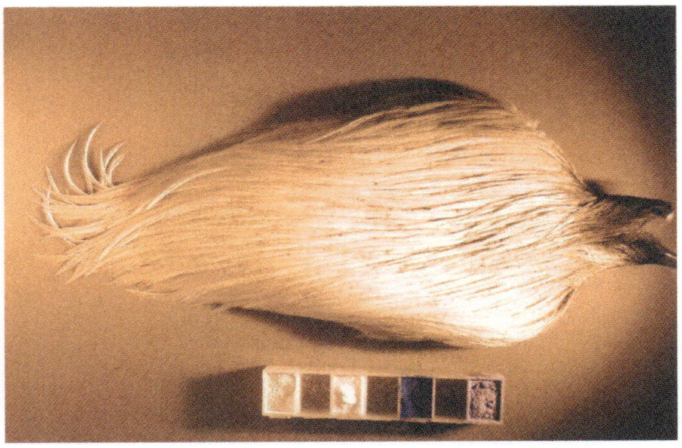

and this white neck is 90%+ UVR.

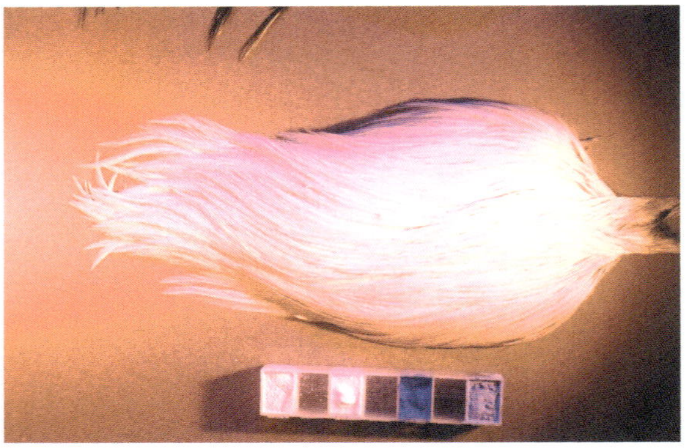

To verify that the UVR comes from the natural feather and not the tanning process, I took photos of the hide side of the white neck, which would have equally absorbed any tanning chemicals. You will note that the skin itself is only average UVR, thus indicating that the feathers' extremely high UVR is an attribute of the feathers themselves, not chemical tanning.

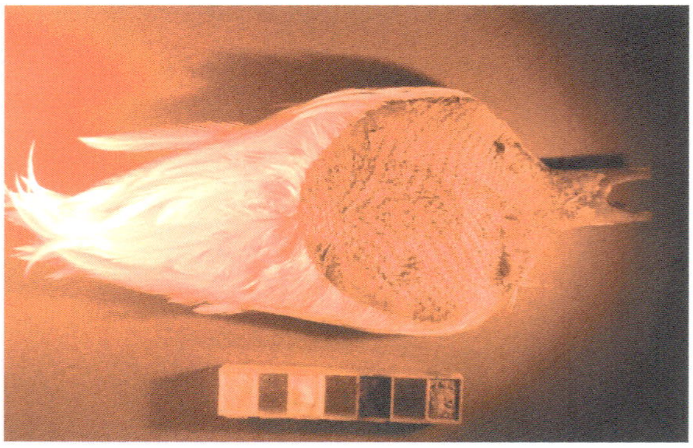

It is difficult to account for the difference in UV markings on these birds; obviously it was not a survival adaptation. Perhaps, as the birds were being bred over many generations specifically for optimum plumage, the crossing of certain genes tweaked the gene for UV marks. Since the science indicates that UV reflecting feathers are generally a property of the structure of the feather and not a pigment,[19] it seems that a fortunate happenstance gives us these UV-reflective natural hackles.

All of the necks shown above are natural, i.e., un-dyed, Metz necks. Natural colors were always highly prized among both fly tiers and anglers. "Natural Blue Dun" were once words to conjure with. Innumerable stories of the search for the perfect Blue Dun cape may be found in fly fishing's prolific literature. In *Trout Fishing in the Catskills* by Ed Van Put (Skyhorse Publishing, NY, NY, 2007) we read of Harry Darbee, the famous Catskill area fly tier:

> They [the Darbees] refused to use dyed or bleached hackle when they tied Catskill-style dry flies; nor would they use furs that were tanned, preferring raw furs with the natural oils left in the hides. Harry was a big believer in natural blue dun hackle, and he was convinced that flies tied with natural dun, with its various tones and hues, reflected light in a unique way and therefore better deceived the fish. The Darbees always kept chickens and raised a large number of dun roosters of various shades, always trying to improve the quality of the hackle.

In 1969 Art Flick wrote:

> "The dressings of the first three dry flies mentioned in this book call for natural dun hackles, all being the same in that respect.
>
> Too much stress cannot be put on the fact that natural dun hackles are far superior to those that have been dyed a dun color.
>
> It is unfortunate for fishermen that the supply of these hackles is so limited. Were they available, there would of course be no necessity for dyeing white hackles to try to imitate the natural dun color, a characteristic color of the wings of so many May (sic) flies. There is no comparison between the natural and the dyed, although most fly-tyers hate to admit it.
>
> Some writers recommend rusty dun hackles for some flies, medium duns for others, and dark duns for still others. As nearly as I have been able to determine, the exact shade of dun is of little consequence. It has been my experience that in imitating any fly that has a true blue-dun-colored wing, the exact shade of dun hackle used is relatively unimportant.
>
> In checking over hundreds of specimens of the naturals, I have found there is a great deal of variation in the shade of the wings. Those on some flies of a species are much darker than the wings of others of the same species.
>
> With this difference in shades occurring in the naturals, there seems to be little need to worry about the slight variation of color in the hackles of the artificial. I have found that for all practical purposes, a dun hackle of medium shade works equally well for (sic) Quill Gordon, Hendrickson, and Red Quill.

Art Flick's New Streamside Guide (Crown, NY, NY, 1969)

We no longer have difficulty obtaining natural blue dun hackles. Harry Darbee sent eggs from his birds to Andy Miner, who added a British bloodline to create what Darbee considered the finest capes in the world. Andy was also generous with his stock and sent eggs to many of the early and great names in the hackle trade—Quarles, Wetzel, Collins, Hebert, and Metz are among them. Each of these breeders then found ways to improve the breed to produce hackles of better length, luster, barb count, and usability. The images above are all of modern Metz capes.

Dr. Tom Whiting bought a separate line of birds—the well-respected Hoffman line—and started further improving the quality. In 1997 Whiting purchased the Hebert line, which had its roots in Miner and Darbee. Dr. Whiting's skill and patience enabled him to cross the Hebert stock genes with Hoffmann grizzly birds and also a white line that he had established, to create the Whiting "Hebert-Miner" natural Blue Dun capes that are readily available today.

Natural Light Blue Dun hackles were widely used on both dry and wet flies. There was something about them that just seemed to catch trout. Could it be the very high UVR? The alternative to natural blue dun in the 1930s-1970s was a dyed product created by applying silver nitrate to white capes. Not only was the color not fast—the hackle turned pink when exposed to the sun for long periods—but it is my belief that the silver nitrate would have produced a feather with almost zero UVR. I say this because of the low UVR of silver (Appendix B) and the fact that silver nitrate is altered by UV.

However, modern gray dyes applied to white feathers often provide strong UVR. Below is a gray-dyed goose quill beside the natural light blue dun cape.

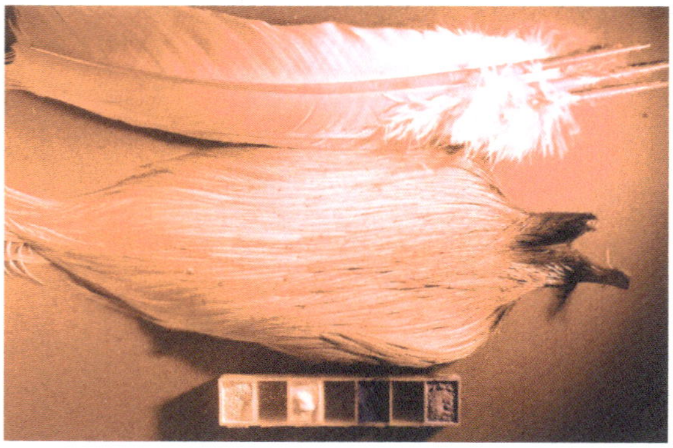

Dyes sometimes impart, either through the dye-stuff or the mordant, a higher UVR than is present with the natural feather—as you will note from the series of photographs of Marabou to follow. Since this UVR is produced by pigments rather than the natural microstructure, it can be reproduced consistently.

Domestic Feathers – Ducks and Turkeys

The agricultural industry has provided us with several sources of highly UV reflective feathers—white ducks and white turkeys. "...the feathers of domestic fowl strongly reflect UVA (Prescott and Wathes, 1999) and markings in the plumage of turkeys (Sherwin and Devereux, 1999) have been found to be visible only under UVA radiation."[20][21]

The white feathers below are white goose shoulder on the left and white turkey biot on the right. The goose seems whiter in visible light, but in reflected UV, the turkey far outshines it. "All that glitters is not gold…"

The images below indicate that the effect of the lower reflectivity of the white goose shoulder does not carry over to the dyed specimens. The pink in the bottom row outshines the natural white on the far right. Of equal interest is the fluorescent yellow in the bottom row which has very high UVR while the similar regular yellow in the middle row is UV absorbent.

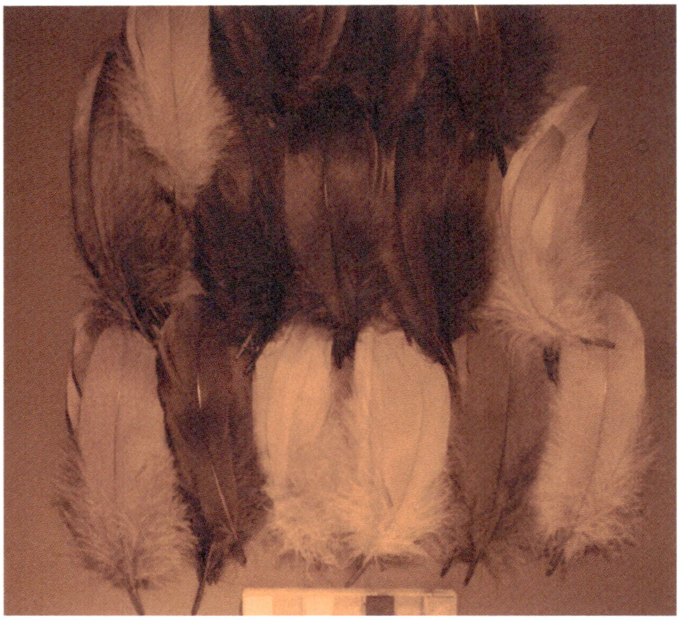

Turkey Marabou

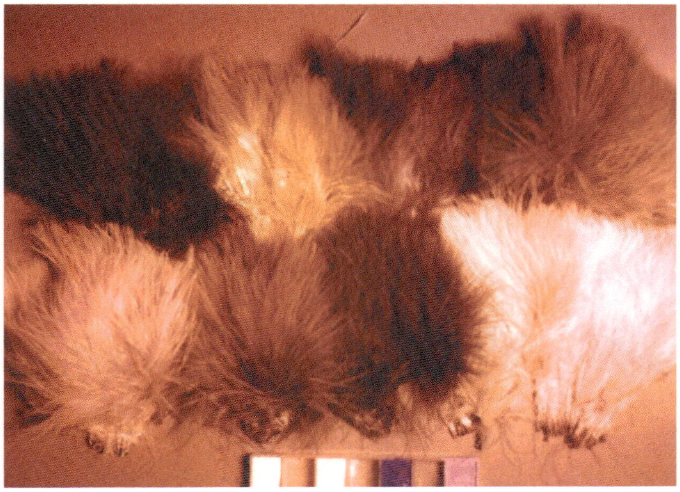

Note the extremes of UVR: yellow (top row, second from right) doesn't reflect even a discernible amount, while fluorescent yellow (top row, fourth from right) has moderate UVR. White and gray are very high UVR, while hot pink is slightly less intense. Fortunately, the supplier of these feathers has documented the results above, and more, and will make that information available. Unfortunately, we can't extrapolate this data to any other supplier because we don't know their dyes or processing methods.

The Marabou stork of Africa was the source of our early marabou feathers, now we substitute turkey down. The name comes from the *marabout*—the term for a Muslim dervish of Africa thought to have supernatural powers.

Wild Duck, Pheasant, and Turkey

The images below indicate that there is UV reflectance from the white markings of green-winged teal and mallard, the tips of turkey tail feathers, mottled turkey quills, pheasant primaries, and the light gray of pheasant wings. The markings are strong enough to merit their use where low to medium UV is required.

You will notice that the right wing in each pair of duck wings is smaller in size than the left. A friend, and superb fly-tier, pointed out to me this interesting fact: when he needed duck quills he could always find plenty of usable quills on the left wing, but few on the right. Immediately I understood the complete global implications of this information—ducks don't actually migrate, but their stronger left wing causes them to fly in very large circles, of which we only detect the resting points.

- Overmywaders

Mallard wings, especially the white-tipped iridescent feathers, are a fly-tiers staple. The white has a very strong UVR.

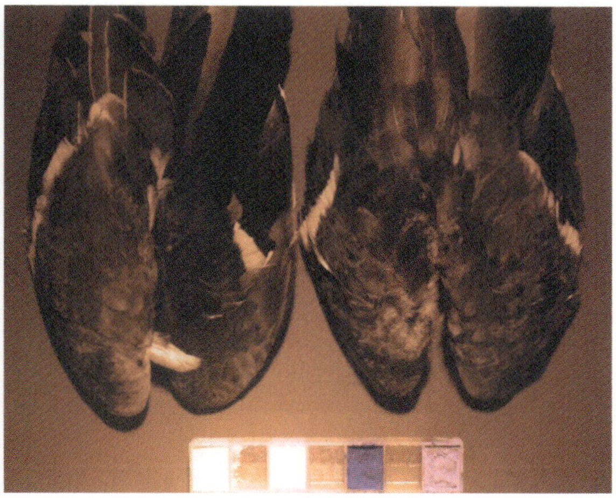

The gray mallard secondary feathers are used for many wet and dry fly wings, for example the Black Gnat, Leadwing Coachman, Iron Blue Dun, and Henryville Special. Surprisingly, the feathers show a very strong UVR, especially in the lighter sections of each feather.

Duck flank feathers have always made good wing material in trout flies. Below are Gadwall, Pintail, Wood Duck, and Mallard flank feathers.

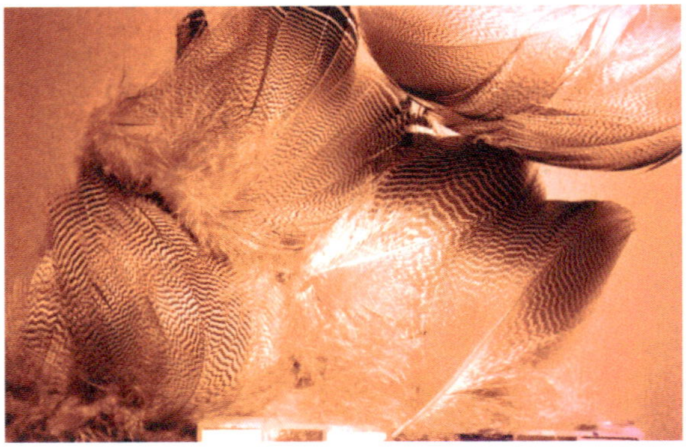

You may be able to detect the UVR, but this is one of those cases where a filter mask of the UV shot over the VIS shot helps, such as the image below.

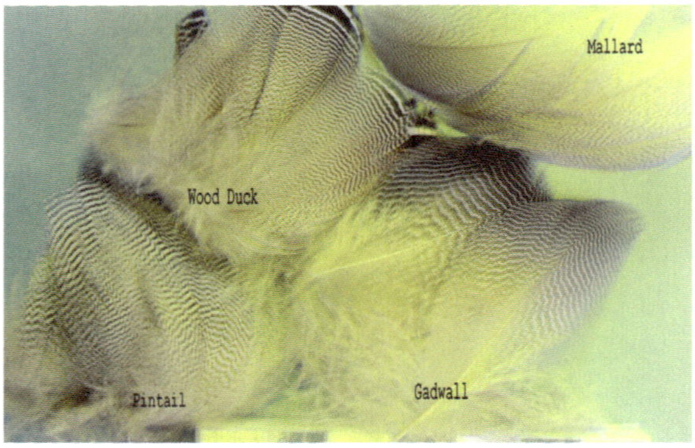

The feathers in the image above are, starting at the top left and proceeding clockwise, Wood Duck, Mallard, Gadwall, and Pintail. The yellow pseudo-color highlights the areas of greatest UVR. The Wood Duck, Mallard, and Pintail have good overall UVR, the Gadwall has less toward the feather tips.

The Ringneck Pheasant and Golden Pheasant skins shown below have quite a few medium UVR patches of feathers.

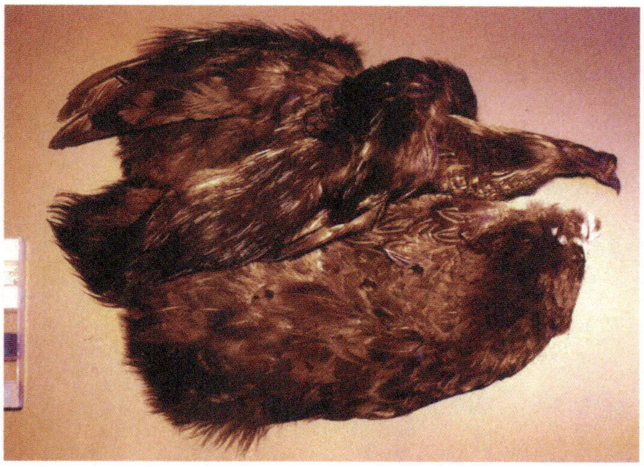

Some of the biots and quills of wild turkeys have strong UVR markings. The images below illustrate a good cross-section of wild turkey feathers used in fly patterns.

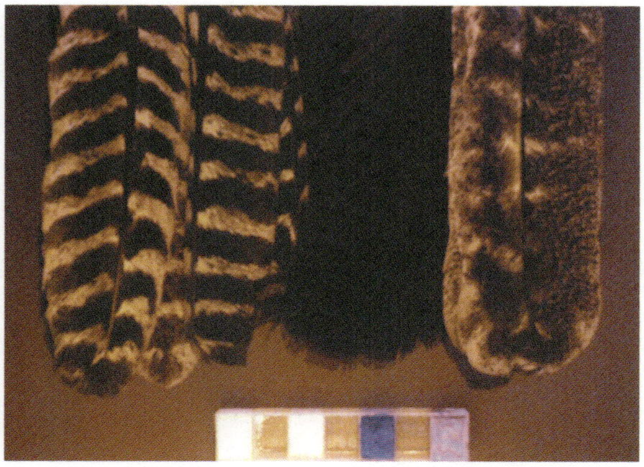

Feathers from other wild birds

With the advent of the CITE listing[35] which prohibits the possession or sale of many feathers and skins once popular in fly tying, e.g., blue chatterer, wild junglecock, condor quills, and blue jay, the fact that most songbirds carry UV markings is not relevant to the fly tier, though interesting.

However, certain feathers—now usually procured from farm-raised birds—are still widely used; for example, emu, ostrich, peacock, and guinea hen.

Emu, Ostrich, Silver Pheasant, and Guinea Hen

Note the high UV reflectance of the Silver Pheasant and Guinea Hen. The more moderate reflectance of the Ostrich is still greater than 31%.

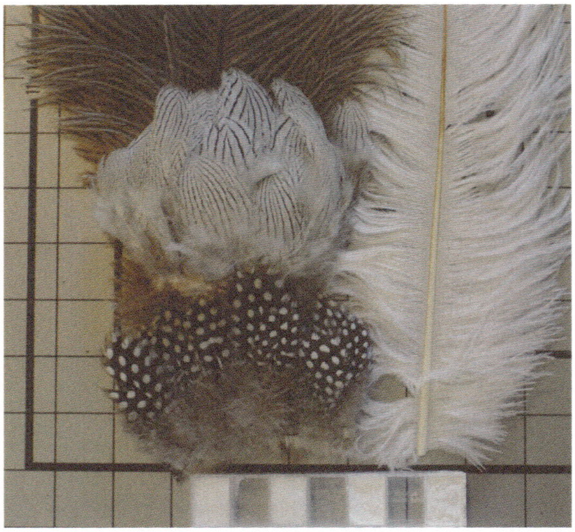

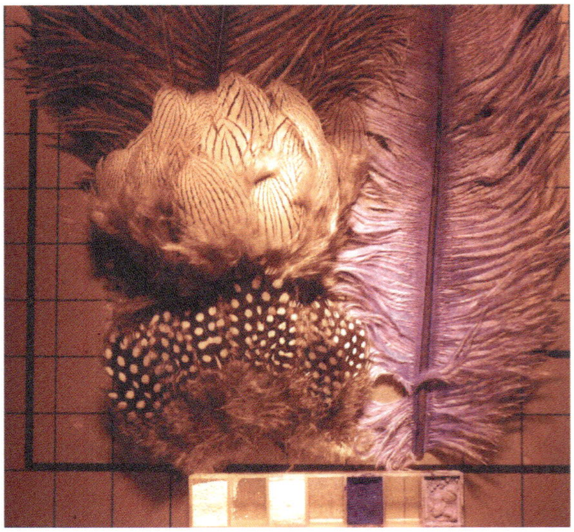

Iridescent Feathers

The iridescent materials below would appear very dark in deep or dark water. When used in combination with a highly UV reflective material, the resulting fly would be a solid target in murky water or on dark days.

Dyed Quills

Dyed duck, goose, and turkey quills and biots make excellent winging for traditional and modern patterns of all types, both wet and dry. Not only are these feathers dyed the strong colors of the marabou previously shown, but the subtlest shades of olive, tan, and gray as well. It is unnecessary to present all of the possible dye patterns in this book, but an indication of how subtle changes in color can affect the UV reflectance is certainly in order.

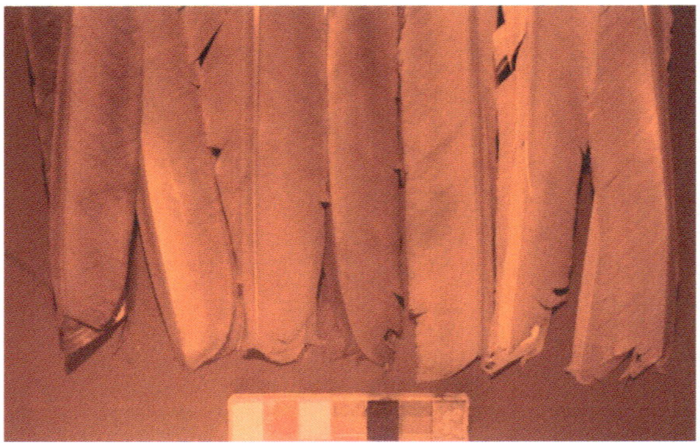

You will notice that, with the exception of the medium gray, the reflectance is quite similar, right to left, until you reach the olive on the far left, which is slightly darker. All of the quills above have a significant UV signature that can be used when tying.

A group of less subtle dyes shows the following results:

The mahogany biot, second from right, and the dark brown, at the far left, are highly polished and show a specular, rather than a diffuse reflection. The sulphur orange, third from the right, has the strongest diffuse UV reflectance of the group.

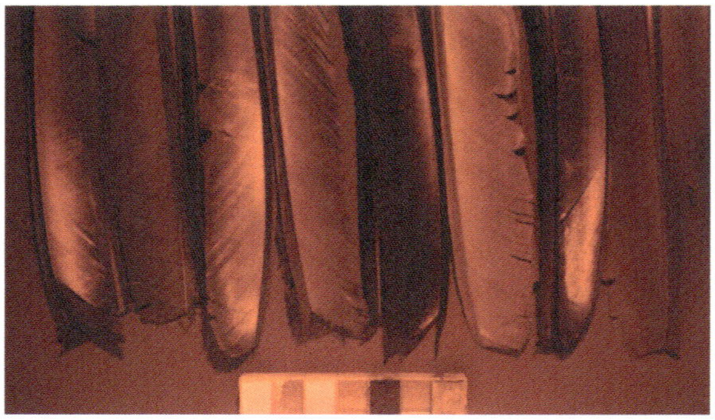

Furs and Hairs

Some other natural materials we commonly use in our mayfly imitations—furs and hairs—do not necessarily provide a UV key, the all-important "trigger," which tells the wary trout that the fly drifting above him is real. Fur-bearing animals, from whom we get much of our fly-tying material (let's not question their generosity), can find potential mates through VIS vision, sound or scent, so they do not require UV vision or UV markings for this purpose. UV markings provide no advantage for most carnivorous mammals in hunting—stealth would dictate camouflage and concealment. Prey animals equally would not benefit from markings that reflect light in the UV to any potential UV capable predators or prey. The "flag" of a whitetail deer—the underside of the brown tail—is strongly UV reflective but protected from view of predators until it is raised in alarm to warn the rest of the herd.

Other herbivores and omnivores do have UV signals. The tail of the grey squirrel has bright white markings which reflect UV.

Perhaps these markings allow quiet signaling in low light. (I've never been able to determine what those critters are saying with their tails, but it is sarcastic… count on it.)

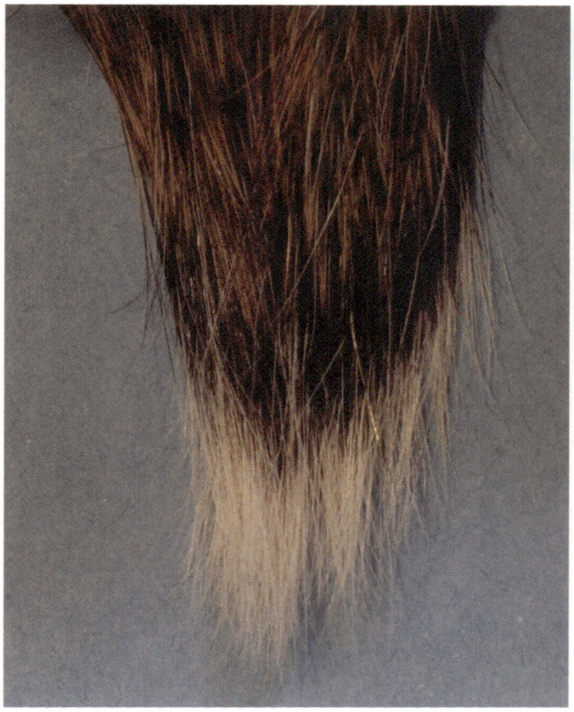

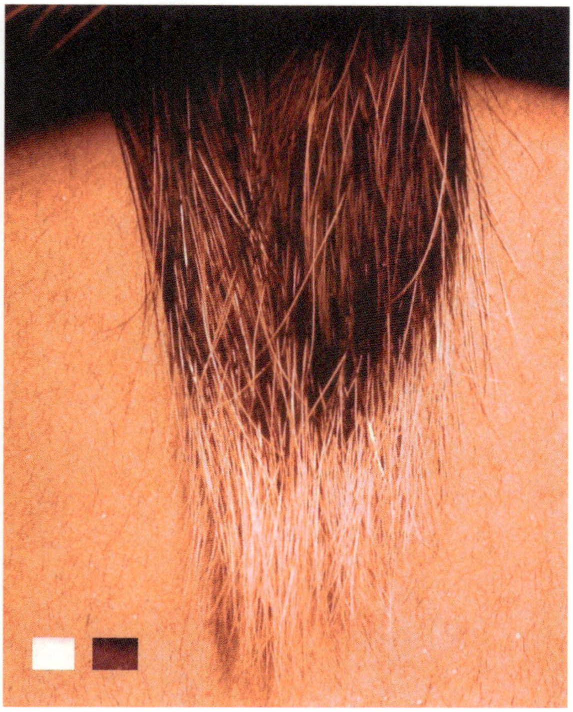

It was once believed that the translucent guard hairs of polar bears acted like optical fibers, channeling the UV to the black skin of the bear, thus providing heat. However, recent studies reveal that the hair absorbs little of the UV;[27] logically, if the hair absorbs little UV, then it must, by extension, reflect much.

For every rule there are numerous exceptions. Snow reflects UVA at 85% (see Appendix B). Animals, whether predator or prey, which hunt or forage during long periods of snow are often white, to blend with the environment. But being white is not enough; they should also reflect UV at the same rate as the snow or they will stand out to their UV-enabled predators or prey.[36] Hence, white snowshoe rabbit fur naturally reflects UV. Arctic fox, as shown by the specimen of fur below, has high UVR, approximately 80%+.

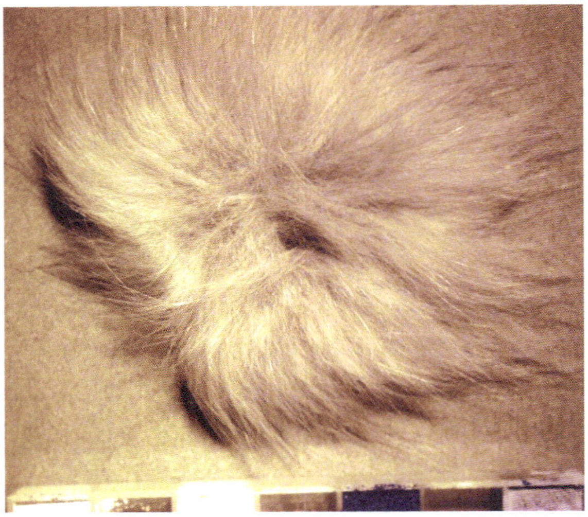

Following this line of speculation, Polar bear and ermine probably exhibit a UVR similar to Arctic Fox and snow. Below is natural snowshoe rabbit (Varying Hare). This particular specimen was caught between light and dark phases; nevertheless, the white areas of fur are highly UVR.

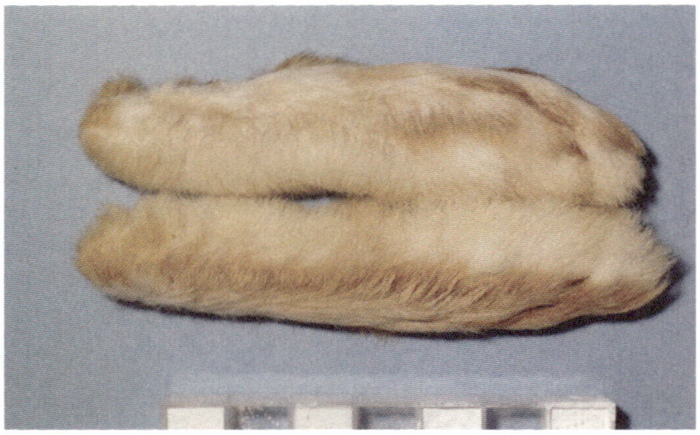

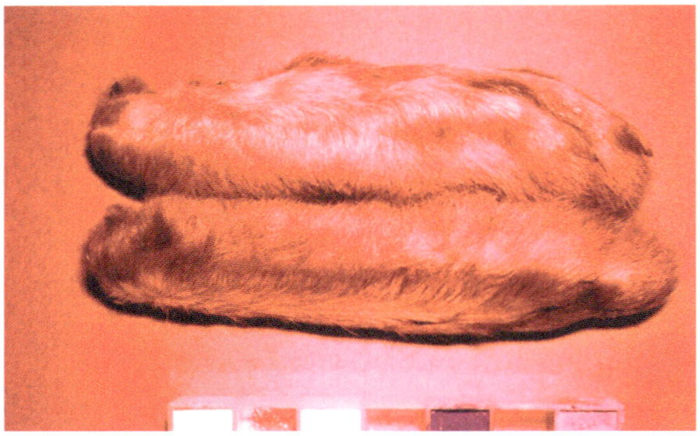

Why white moose mane reflects well in the UV is a mystery to me. I suppose when you are that large you can wear red to a funeral and no one would dare say "No."

Below is a popular dubbing fur, muskrat. As you will note, it reflects very little UVA, almost in the 3% range. This is very useful to the fly tier for simulating the bodies of those insects which reflect little UV. We should consider UVR in all its intensities when designing flies.

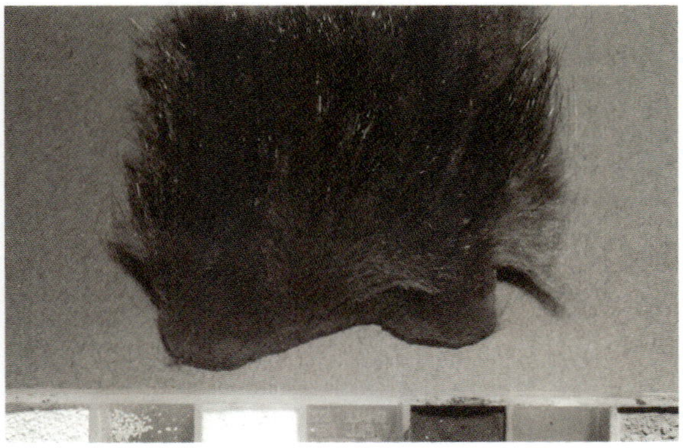

While most natural furs and hairs may not reflect strongly in the UV, application of dyes, sometimes applied with "setting" chemicals called mordants, may render the fur highly UV reflective.

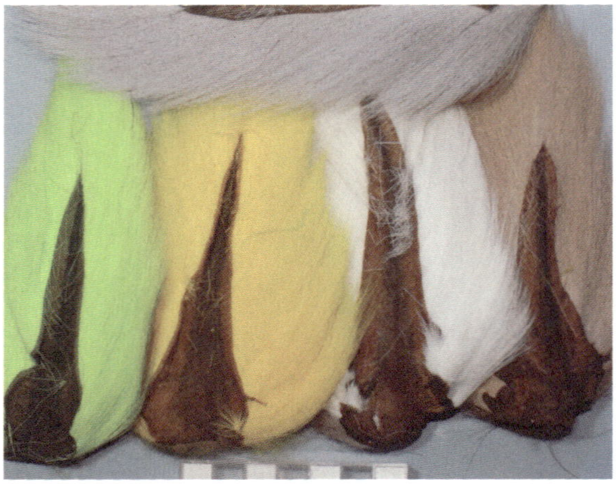

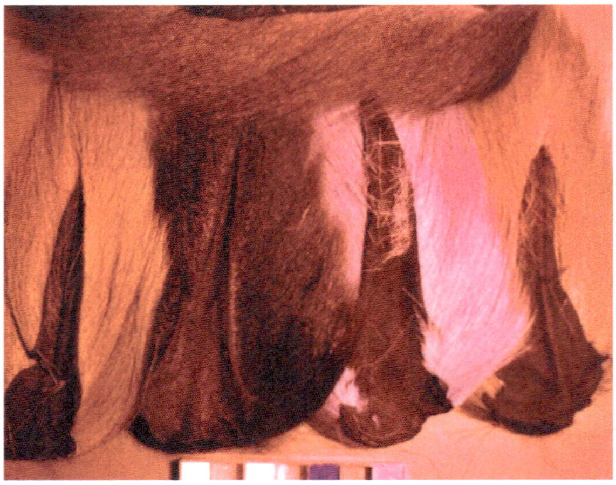

The white bucktail above is highly UVR. The Tan bucktail at the far right and the Chartreuse at the far left both have a medium UVR. Note how the yellow bucktail, second from the left, like the yellow marabou treated by the same firm, has almost no UVR. However, hairs take dyes differently than feathers, so do not expect all yellows to be low in UVR. This similarity seems coincidental—perhaps a common pigment was used.

Rabbit Hair Dubbing – Hareline Dubbin

Rabbit hair has always been popular as dubbing because of its ease of use and availability. Below are some of the popular colors of rabbit hare dubbing. Note the intense UVR of some of the colors.

The colors represented are, top row, left to right: Hare's Ear, Brown; middle row, left to right: Olive Dun, Light Gray, White, Cream; bottom row, left to right: March Brown, Olive, Light Cahill, Golden Brown.

Other Natural Dubbing – Superfine and Micro Fine

The Superfine and Micro Fine Dubbings are very popular with tiers of dry flies. Below is an assortment of popular colors.

Top row, left to right: Sulphur Yellow, Brown, Tan, Baetis Tan; bottom row, left to right: PMD Olive Dun, Light Cahill, PMD Yellow Olive, Cinnamon. You'll notice that the Tan, though darker in the VIS than the Sulphur Yellow, is very high in UVR.

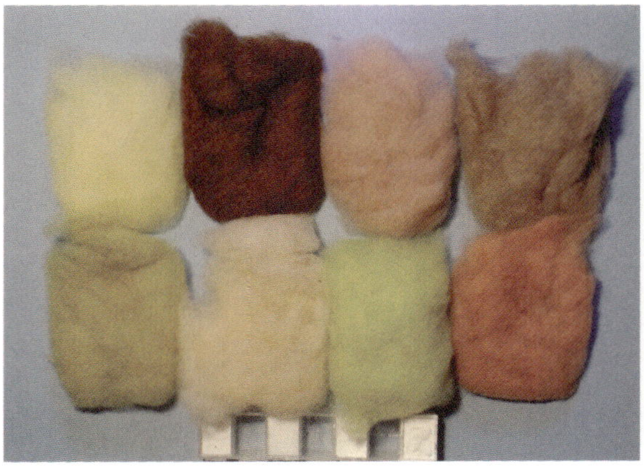

CHAPTER TWELVE

SYNTHETIC FLY TYING MATERIALS

There are so many synthetic materials in use in fly-tying that it would take a lifetime to chronicle the UVR of them all. Fortunately, at least one major supplier will be indicating the UVR of its products this year.

A précis of a few popular items follows.

Ice Wing/Ice Dubbin

Hareline Ice Wing is a polyester-based Ice Dubbin in 8" lengths. The photos above would be applicable to both products.

Note that the gray on the lower right has the highest UVR.

Trilobal Dubbing

The following images represent a popular Antron-based dubbing, Hareline Trilobal.

The colors are, top row, left to right: Dark Olive, Tan, White, Cinnamon Caddis.

Bottom row, left to right: Damsel Green, Seal Cream, Brown Stone, Gray.

Here the gray on the lower right, though very similar in color to the Ice Dubbing, has low UVR rather than high.

White Antron, here as dubbing but also available in longer strands, is a popular post for parachute dry flies. Considering its very high UVR, this is doubtless an excellent choice.

Synthetic Floss

The spools of floss in the images below are from many manufacturers and sources. As you will note, flosses, like many of the materials we have examined, are richly diverse in their UVR. A yellow from one maker may glow brightly, from another the same yellow will have little UVR.

Floss in UV converted to grayscale for greater clarity:

The bright bars of light on each spool in the UV photo are from specular reflectance. To appreciate the diffuse reflective property of the floss, look at the relative lightness of the band beside the bars on the spool. From the top: you will note that the green, pink, and gray on the top row are the most reflective in that row; in the second row it is—from left—the yellow and the pink; the third row—the white , pink, and blue. None of this is intuitive; the dyes determine the UV reflectance of the floss without an association to the visible color. For example, the green on the first row is highly UV reflective, but a very similar green on the second row is not. The grayscale image, made from the UVR image, may provide a clearer reference.

Thread

The color of the tying thread is usually specified as an element of a fly pattern. This becomes especially important when UVR is considered, as the UVR can pick up highly reflective thread glowing through a dubbing body. The threads in the images below are all Uni-Thread from Uni Products, Inc. As you can see, use of the white, cream, and pale yellow will provide a significant UVR spot on a small dry or wet fly, especially at the exposed head.

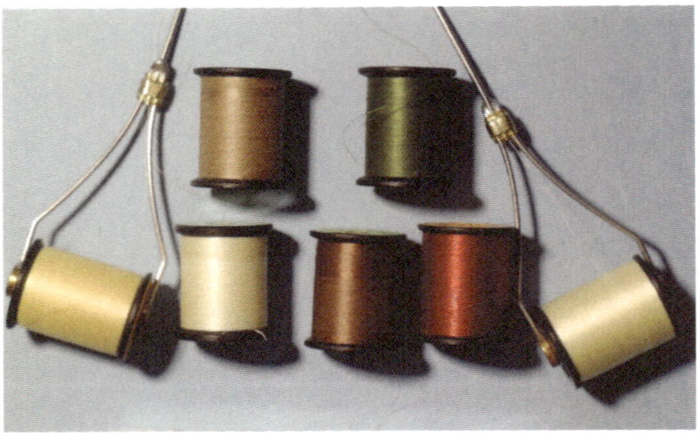

"Castera was every way unequal to his task. He did not perceive his author's beauties. He either suppresses or lowers the most poetical passages, and substitutes French tinsel and impertinence in their place."

- *Dissertation On The Lusiad, And On Epic Poetry*
By William Julius Mickle

Tinsel

Tinsel is a standard component of subsurface flies where it serves to add specular flash to the body of the fly. The ragged collection of tinsel in the images above is from many dates and sources.

Any diffuse UVR is found between the bands of light in the UVR image above right. You will notice that the Opalescent Mylar tinsel (top row, second from left) has the strongest UVR; second to that is the silver tinsel directly below it, which is also made of Mylar. The two gold tinsels and all the colored and textured tinsels are very low in UVR. The three other silver tinsels are medium UVR and are all metallic. It may be safe to infer from this that if you wish to add flash to a fly with tinsel and have it reflect well with both diffuse and specular reflection in the UV, Mylar tinsel is a good choice.

In the photos below we have what is known as "French Tinsel." French Tinsel has been used for centuries in fly tying because it is less susceptible to corrosion than most other metallic tinsels. French Tinsel actually became a pejorative in Britain as early as 1780; indicating speech or writing with flash but no substance.

However, there is plenty of the flash that we can appreciate. Notice the strong UVR of the silver tinsel. This would make an excellent body or ribbing, providing both specular and diffuse UVR.

CHAPTER THIRTEEN

OTHER TACKLE

Fly Lines

Fly fishermen have traditionally used dark lines when fishing subsurface and light colored lines when fishing dries. It is much easier to follow the drift of small flies if you can see the line clearly. The popular colors for floating lines have covered the spectrum from white through blues, greens, tan, pale gray, to fluorescent orange. The dry fly fisherman tries to avoid "lining" the trout, i.e., allowing the line to pass over the trout's position, so the color of floating fly lines has always been deemed insignificant to the taking of fish.

Taking the line's UVR into consideration would seem immaterial if the angler is perfect in not lining fish. For those of us who might occasionally line a good trout, a line that is neutral in color as well as medium in UVR is optimum. Moderation is the key; too much UVR might startle the fish, as would, probably no UVR. The same might be said for sinking fly lines.

The UVR of the lines pictured below was remarkably consistent. The Wulff Olive had the lowest UVR, the SA Sharkskin Chartreuse had the highest, but all of the lines were reflecting in the range of 30% plus or minus ten. This is encouraging as it is neither so low as to stand out in a light UVR background, nor so high as to attract attention as a glowing strip.

One fly line, not of this set, a Scientific Anglers "Concept" fly line which is sold as a beginner's line for approximately $12 US, is worthy of note. The UVR image below shows the almost total UV absorption of the line. Line a trout with this and it is unlikely you will take him on subsequent casts. A further consideration is that since this line absorbs UV so readily, the plasticizers in it will probably break down quickly, reducing its life span.

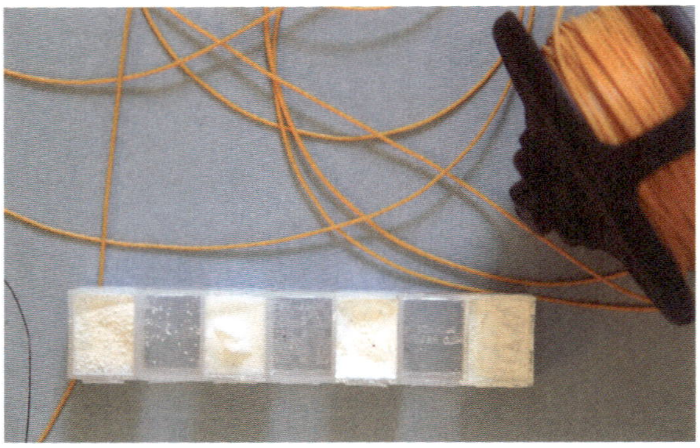

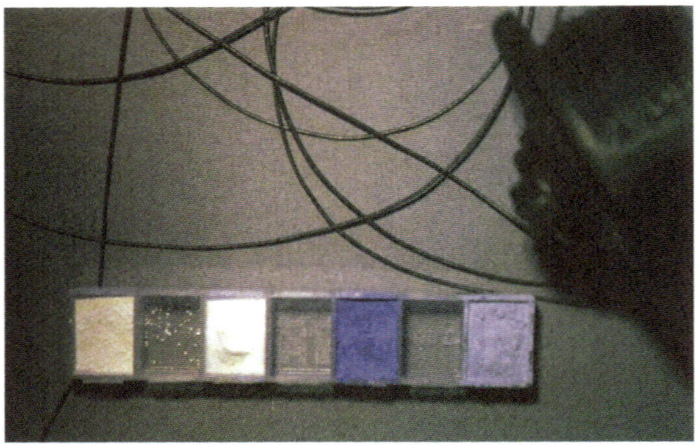

The "x" designation in leaders and tippets harks back to the time of drawn gut. The silk sac of silkworm caterpillars was pulled out, after pickling the worm. The silk hardened as it hit the air. This produced a single strand of "gut" seldom more than 18" long. The gut was then drawn through a die—a hole in a diamond of precise size. The first time it was drawn it was called 1x. It might then pass through a second die .011" smaller, shaving it to 2x, and so on. The term "x" follows the size measurements precisely still, though the Nylon is extruded rather than drawn.

Leaders

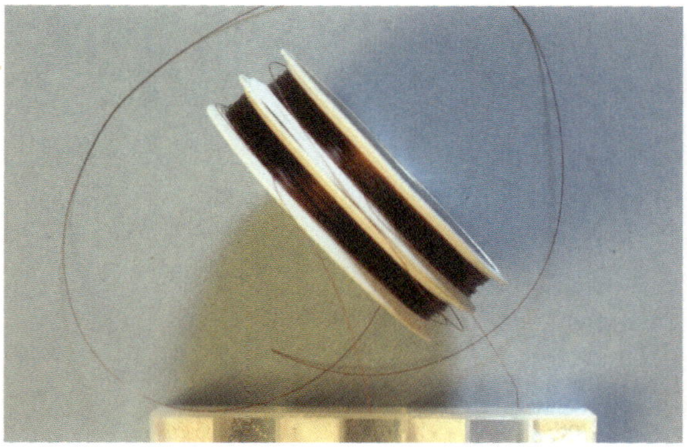

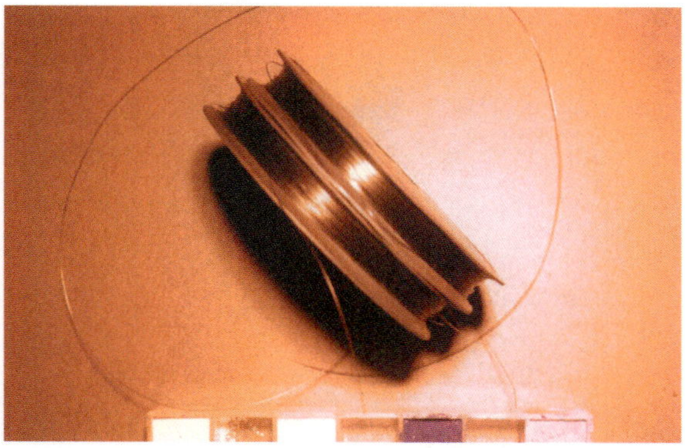

Maxima Chameleon

The leader material above is 1X and 3X, which is quite stout if it were used as a tippet. Nevertheless, despite the brown-green color in VIS, the UVR is a pleasant neutral, which should cause it to blend well with most ambient UV conditions. The closeup (right) emphasizes this.

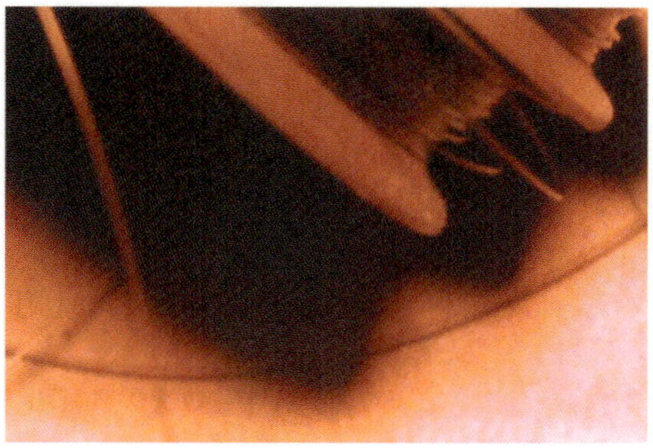

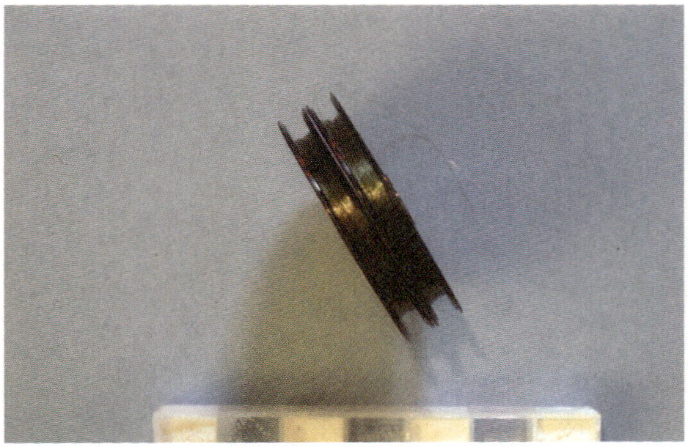

Orvis Super Strong

Orvis Super Strong tippet material also has a UVR that blends with the neutral UVR background. To the right is a closeup of the Super Strong. This material would probably be an excellent choice for UVR invisibility.

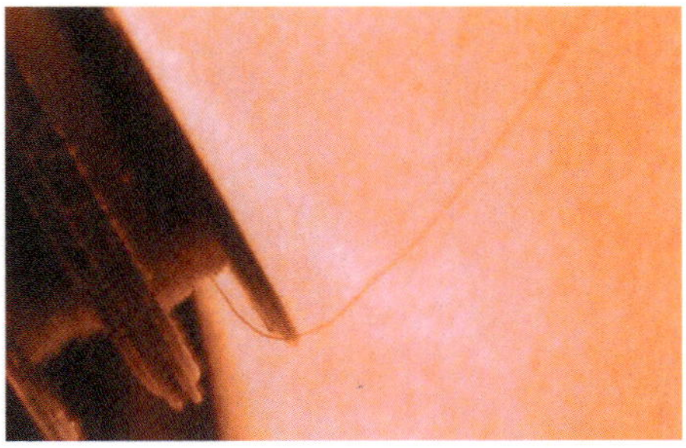

Orvis Knotted Leaders

These Orvis Knotted Leaders are not new. In fact, I think they are from the 1970s, which shows how often I change my leader. The modern Orvis Super Strong 7X is 2.5 lb test, not 1.0 lbs of these leaders. However, the fine use of green tint and medium UVR makes them good leaders even today.

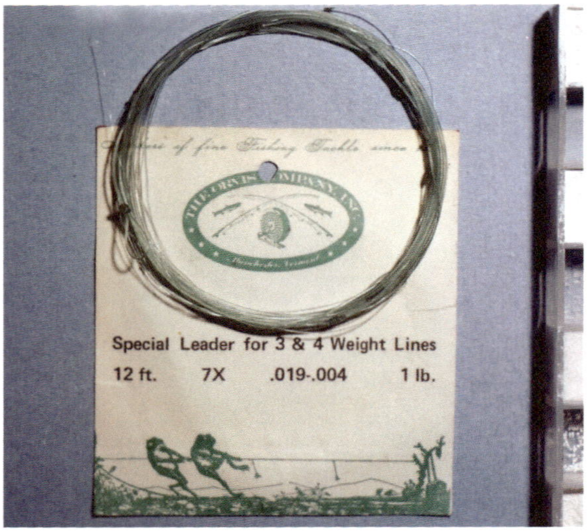

Clothing

Below is an example of typical streamside summer foliage in New England. The photo is taken with the sun ahead at approximately 70 degrees above the horizon.

In the UVR image below, the water appears lighter than normal because of specular reflection from the surface. While water reflects only 5% of UVA striking at a 90 degree angle to the surface, specular reflection will still hold true, causing a greater amount of reflection at 70 degrees. The flowering plants are medium UVR in this photo while the leaves of the trees absorb UVA. This is an average background for many trout fishermen.

How does this affect our fishing? Well, if we are intent on remaining hidden from the trout by blending into the background, we should make sure that our clothing blends in both VIS and UVR. Here is a typical khaki-colored fishing vest in VIS and UV.

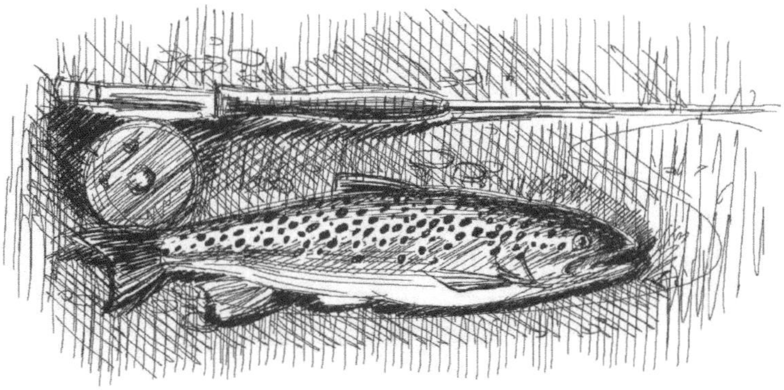

CHAPTER FOURTEEN

SOME ANECDOTES TO ILLUSTRATE UVR IN FISHING

The case for Bivisibles

From *The Practical Fly Fisherman* by A.J. McClane (Prentice-Hall, NY, 1953)

"To be a dry fly fisherman all you will need in the beginning are four flies: a bivisible, a spider, a bunched wing pattern, and a fly with white wings that you can see at dusk. You will want more than four patterns of course, but the secret of success in floating flies is to be ready with the type of artificial that suggests an insect form and at the same time fits the water you are fishing. From the basic four you can build a box based on your experience and the experiments of other. The value of certain patterns may change from one stream to the next, or even one day to the next, but as a practical angler you will want to give your future some direction.

Charles Ritz, a man with over thirty years of experience, represents the "functional" school. He uses a series of bivisible in pink, gray, white, black, brown, and yellow, all designed for visibility rather than imitation. The colors are selected according to the prevailing light conditions. When the sun is angular and fairly burning into his eyes, the dark, black bivisible is by far the easiest to see. In calm water with a strong back lighting, a pale colored fly is best. Intermediate shades, such as pink and yellow, fit in when strong reflections from bank foliage create a changing surface for the fly to drift over. Each color, or pattern is tied in two different styles—one sparse for slick water, and rising fish, and the other tied full, to bounce on fast water as an attractor. Perhaps Charles is an unkind realist in this simplification, but he's one of the craftiest fly-fishermen who ever duped a trout.

We were fishing a broad stretch of the Neversink one evening.

The flat was more crowded than usual—and nobody had been taking fish. A hatch of flies came over the water about dusk, and up at the head of the flat I saw Charley lay out a long cast and hook a heavy brown who kicked up quite a rumpus. The fellow fishing below him asked what fly he was using and Charles allowed that it was a Coachman. In a moment the word that binds all anglers sounded like drums along the Mohawk—everybody was passing the information on. There was a clicking of fly boxes and some streamside swapping, and I know for a fact that two people fishing just below me caught trout soon afterward. I fished for a while and finally walked up in time to see Charley net his third trout. While admiring his catch I noticed the fly in the trout's jaw—it was a scraggly white bivisible. He avoided my glance and watched the shapeless figures shooting their lines at the growing darkness. We didn't speak for several minutes, until one angler let out a whoop of victory; then Charley remarked quite casually, "Everybody has faith in the Coachman." "

In the instance above, Charley Ritz used only Bivisibles, which we have found have a very bright UVR signature in the white front hackle. The fly he recommended for the other fishermen to use, the Coachman dry, has a bright UVR signature in the white quill wings. They were fishing at dusk, where UV prevails. Is it possible that Ritz caught fish consistently on his various Bivisibles because of the white hackle providing the necessary UVR marking?

The great mystery of the Royal Coachman

Anglers have been attempting to explain for generations the unlikely effectiveness of the Royal Coachman—as wet fly, as quill-winged dry, as fanwing, and as hairwing. Preston Jennings, in his superb work *A Book of Trout Flies* (Crown Publishers, NY, 1935) mentions the Royal Coachman five times. Jennings was a strict imitationist when it came to trout flies; it irked him that the Royal Coachman—resembling no known insect life—consistently caught fish. In 1956, Jennings felt he had found an answer to this enigma and published an article in Esquire Magazine entitled "There *IS* a Royal Coachman," explaining that the fly imitated the *Isonycia bicolor* nymph. Arnold Gingrich enthusiastically spent a chapter of his "Well-Tempered Angler" extolling Jenning's theory. Art Flick, another fine angler and author of *New Streamside Guide to Naturals and Their Imitations* (Crown, NY, 1969) was flummoxed by the Royal Coachman's enduring effectiveness. Over several pages he struggled for an explanation, but came up short.

I will now be so bold as to suggest that—as the photos (right) of the Royal Coachman in Hairwing dry, Fanwing dry, and quill wet illustrate—the white, highly-UVR wing in conjunction with the UV absorbent peacock herl provides all the visibility and feeding triggers that a trout needs. The key, then, to the perennial success of the Royal Coachman series of flies is its UV signature. The same might be said for the Coachman, Prince Nymph, and Zug Bug—all very "taking" patterns, all sharing certain UV characteristics.

Fly design is an art understood only by the depraved and the writers of Monty Python. In the famous Pet Shop Sketch—no not the Dead Parrot sketch, the other one—the discerning artists of MP illustrate proper design techniques.

Caption: A PET SHOP SOMEWHERE NEAR MELTON MOWBRAY

CUSTOMER Good morning, I'd like to buy a cat.
SALESMAN Certainly sir. I've got a lovely terrier. [indicates a box on the counter]
CUSTOMER No, I want a cat really.
SALESMAN [taking box off counter and then putting it back on counter as if it is a different box] Oh yeah, how about that?
CUSTOMER [looking in box] No, that's the terrier.
SALESMAN Well, it's as near as dammit.
CUSTOMER Well what do you mean? I want a cat.
SALESMAN Listen, tell you what. I'll file its legs down a bit, take its snout out, stick a few wires through its cheeks. There you are, a lovely pussy cat.
CUSTOMER It's not a proper cat.
SALESMAN What do you mean?
CUSTOMER Well it wouldn't miaow.
SALESMAN Well it would howl a bit.
CUSTOMER No, no, no, no. Er, have you got a parrot?
SALESMAN No, I'm afraid not actually guv, we're fresh out of parrots. I'll tell you what though ... I'll lop its back legs off, make good, strip the fur, stick a couple of wings on and staple on a beak of your own choice. [taking small box and rattling it] No problem. Lovely parrot.
CUSTOMER How long would that take?
SALESMAN Oh, let me see ... er, stripping the fur off, no legs ... [calling] Harry ... can you do a parrot job on this terrier straight away?
HARRY [off-screen] No, I'm still putting a tuck in the Airedale, and then I got the frogs to let out...
SALESMAN Friday?
CUSTOMER No I need it for tomorrow. It's a present.
SALESMAN Oh dear, it's a long job. You see parrot conversion ... Tell you what though, for free, terriers make lovely fish. I mean I could do that for you straight away. Legs off, fins on, stick a little pipe through the back of its neck so it can breathe, bit of gold paint, make good ...

This is my approved technique as well. Say, for example, I want a dry fly that imitates an adult (imago) female mayfly still carrying eggs. The insect has transparent wings; rose highlights in the thorax; a very pale, translucent, cream body with just a hint of red; and a quite noticeable, though pale, pinkish-orange egg sac still internal.

So, I begin with a Light Cahill... well, not really. I begin with pale yellow thread and a tail of pale blue dun hackle fibers, as with the Hendrickson. The body will be that of the Light Cahill, of MicroFine Cahill dubbing (VIS, UVR)

however, I would first tie on a butt of MicroFine Tan (VIS, UVR)

to simulate the egg sac. Wings will be teal flank, upright and divided. The hackle will be natural pale blue dun, ala Hendrickson, but wound over Tan ostrich herl in imitation of the thorax (VIS, UVR).

The head will be the pale yellow tying thread. So, it is exactly like a Light Cahill, mated with a Hendrickson, and pregnant with an alien child.

- Overmywaders

CHAPTER FIFTEEN

DESIGNING YOUR FLIES

Now that we have an understanding of the trout's use of UVR, we can begin implementing that in our fishing. In my opinion, and that of many others, the main point in fooling trout through imitation is meeting the fish's expectation, not achieving perfect realism. As noted earlier, the trout's expectation is a rough pattern to which we must present an acceptably matching *gestalt*. This is why the use of dubbing is so successful: despite the trout's excellent eyesight the broken outline of protruding bits of hair and fur suggest the translucent body of the natural. In using UVR where we are uncertain of the extent of UVR markings, we should apply the same approach—not too much, nor too little, but only enough as to suggest the impression of life.

More is not always better—I learned, for example, that I cannot be trusted at an ice-cream smorgasbord. Even as an adult I will try to apply every single topping to my ice-cream. The resulting sticky mixture is not half as delicious as a few subtle toppings. Just so with the use of UVR: keep it subtle. True, there are splendid opportunities for excess when stripers are "busting bait" and then perhaps the brighter in UVR the better; but for stream and still water trout fishing, these occasions seldom arrive. Applying a small amount of UVR dope to the hackles of a dry fly which has no UVR signature is a safe bet; drenching the entire fly may send your quarry scurrying for cover.

Be sure when designing flies that you do not forsake visible color in order to increase the UVR—the trout will see both. There are enough materials available in all colors and intensities of UVR so that one spectrum doesn't need to conflict with another. For example, let's suppose I want a high visibility yellow that also is strong in the UVR. Many of the yellows we have seen are UV absorbent, but not all. In the chapter above on Domestic Feathers, we saw two examples of yellow goose shoulders: one was highly UV absorbent, the other was highly UVR; yet in VIS the feathers were close in color.

Which side of the balance should we choose? Should we design a lot of UVR into our flies, or just a little? In my opinion that depends upon the use of the fly. With a baitfish imitation I would always have at least one horizontal area of high UVR; in low-light and turbid or high water conditions this would increase its visibility without being "garish." Nymphs would receive a little splash of moderate UVR on the top of the thorax. If the pattern is intended to represent a damselfly or dragonfly, I would use intense UVR for the whole body. While only the male of many species has this marking, I don't think the trout are gender selective when such a large bit of fat and protein drifts by.

If the recipe is for a mayfly which emerges or swarms at dusk or later, then I would use moderate UVR in the body; up to forty percent reflective, but not as a solid pattern, as a broken or segmented pattern. I would use higher UVR in the wings. As we noted in the chapter on insects, the wings of the natural mayflies are heavily UVR marked. Further, the chapter on artificial flies showed that UVR in the wings is the common element of successful mayfly patterns. Personally, I favor higher UVR in the wings than might be true in the natural insect. Also, though this may be my "King Charles' Head;" the bodies of most Catskill-style flies are neither long enough nor substantial enough for a great imitation. Our modern hackles and floatants will buoy a larger body than we commonly use on dries. We should take advantage of this fact and improve our imitations in terms of outline.

For caddis dries, the wings should have low intensity of UVR, corresponding with the plant life on which they are found. Note that vegetation varies in UVR from species to species. However, as we noticed in a previous chapter, most plants fall in the range below ten percent UVR. The bodies of caddis flies require more UVR.

As for hackle, our use of UVR depends on what we consider the hackle to represent in the specific pattern. If it is a palmered pattern, such as some caddis ties, then we can give it moderate to high UVR, as the caddis generally have a brighter underside. If the pattern is to imitate a mayfly and it is a standard upright hackle, then we need to assess whether we think the hackle simulates the thorax, head, and wings of the fly, or the legs. For a hundred years we have assumed that trout take the upright hackle for legs. Why? It seems much more likely that the trout's template allows the hackle to pass muster as wings and thorax. Since many mayflies have low-to-medium UVR in the head and thorax, you might consider that UVR level for the hackle and use a large, higher UVR wing.

Do I always adhere to the design suggestions outlined above? Hardly!

As was noted earlier, dyed tying materials, including hackles, provide endless scope for subtle intensities of UVR within a small VIS tonal range as demonstrated by the Goose Quills in a previous chapter. If the dealers and manufacturers of most of our tying materials follow the lead of Hareline, who will be providing the UVR of their products, we shall have an easy time of using our knowledge of UVR effectively in our tying.

This brings up the question of home dying of feathers and furs. We now know that some dyes are strong in UVR, others medium, and some UV absorbent. However, which dyes are which? This is an opportunity for someone to present us with a breakdown of the common fly tying dyes in both VIS and UV.

And now, the dope…

I mentioned earlier a UVR dope to apply to a fly. I have tested the common petrolatum-based floatants—Gink® and Albolene. Applying these compounds to a Hendrickson dry reduces by approximately 10% the UVR of the fur-dubbing body. However, water has the same effect of increasing the translucence, thus reducing the reflection. The UVR of the hackle is unchanged by the floatant.

Next I tried a popular a desiccating powder that is sold in fly shops as a floatant. I was quite optimistic that this would increase the UVR of the fly. I will let you determine for yourself its efficacy.

First I soaked a #14 Red Quill in water and dried it somewhat with a piece of dry cloth. It was still damp when I put it in the vise and took the following shots of the untreated fly.

Then I used the applicator brush supplied with the desiccant and generously applied the compound to the fly. When I had finished brushing the fly from all sides, the fly and the vise were covered in powder. I blew off the surplus powder by mouth, using all my lung capacity—inspiring an impressive fit of coughing. Blowing off the surplus powder would have been less intense than the first cast of the fly, so I am convinced that I was well within the standard operational parameters of the product. Here are the after shots:

Certainly a damp fur-dubbing body would have retained much more of the UVR powder. Enough was retained in the hackle to slightly increase the UVR in that area and the tail as well. There is obviously some clumping of the desiccant but overall it provided desiccating and UVR enhancing benefits.

I then treated a second #14 Red Quill with "Overmywaders UVR Fly Dusticator" (Patent Pending), a non-toxic, environmentally-safe means of drying the fly and adding UVR. I used the same application method as with the other product. Here are the before photos:

And here are photos of the treated fly:

As you can see, the Dusticator dries the fly very well, separating the hackles and adding a very high UVR, even to the body, without clumping. If you only wanted UVR on the hackle and tail, simply brush it on those areas. It also works for adding UVR to wet flies, nymphs, bucktails, and streamers.

Obviously the previous two paragraphs are a shameless pitch with self-serving intent. However compadres, if you wish to make a good desiccant for your own use, with high UVR properties, and don't mind the labor, trot down to the local pharmacy and purchase some Magnesium Oxide in powder form or in capsules containing powder. Then rush home and run the MgO through a coffee grinder. This will have several positive results:

- You will have a wonderful fly floatant
- Your coffee will have an "interesting" taste
- For a time you will be "regular"
- Your significant other will have an entertaining story highlighting your piscatorial dementia

CONCLUSION

How important is UV vision to the fish? Let's follow the logic:

1. Trout have UV vision
2. Sensory abilities retained in a species are usually significant in sexual selection, food gathering, defense, or all of these; therefore, the UV vision of trout has evolutionary significance
3. UVA light dominates at dusk and dawn
   a. Many insect hatches occur during dusk and dawn
   b. Large trout primarily begin feeding at dusk
   c. Many insects and baitfish have UV markings visible to trout
4. UVA penetrates water well—better/deeper than visible light
   a. UV scattering backlights prey even during high, turbid water conditions, allowing trout to continue to forage effectively

Thus, it would seem that anglers should consider UV reflectance as a very significant factor in their art.

And, the result of application of your new knowledge may be...

Success!

APPENDIX A

How the UVR Images Were Created

Equipment used:

Camera
Nikon D40 digital SLR—the D40 and some of the others in its class are capable of picking up UVA down to 340nm without alteration of the camera. Many digital cameras, including many Nikons, have UVIR filters on the CCD (Charged-couple device—the optical sensor which converts incoming photons to electrical impulses), which must be removed in order to pass UV wavelengths. The D40 creates a 6.7 mega-pixel image, which is more than adequate for modern printing applications.

Lenses
The most serious limitation that UVR photography encounters is the reflection of light from the filters and lenses used. The excellent filter mounted on my camera is intended to allow *only* UVA light to pass. While it performs that function well, it also reflects much of the UV. So, before the UV reaches the first glass element of the camera lens, the amount of transmitted UVA light has been diminished by twenty percent.

If the camera lens uses crown glass, each of the lens elements will diminish the UVA transmission by approximately thirteen percent. This is cumulative. Add coatings and modern bonding agents in lens groups and it gets worse. The sum of the above is that the photographs are working with much less UVA than the trout is getting.

Imagine that the light source is 1000 photons per square millimeter—just an easy number to work with—and it is directly overhead. Nine hundred and fifty (950) photons will penetrate the water's surface and be scattered by particles in the water, thus providing UVA illumination to the trout. A rock in the stream bed reflects 80% of that UVA and the trout receives most of those 760 photons. Now suppose you wish to take a UVR image of the same rock above the water. You start with 1000 photons, 800 photons are reflected by the rock, but only 421 photons reach the camera's sensor. The sensor itself may process from 50-80% of the incoming photons and pass them along to the camera's image processing chip for interpretation. So, the trout eye interprets 760 photons, and the camera interprets from 210-337 photons. That is best case, with a triplet lens and a Baader U2 filter on the camera.

The lenses used were pre-1962 lenses from various German optics firms and two Japanese firms. These lenses were not multi-coated, nor anti-UV coated, and it is my belief that they passed more UVA because of this. Most of the lenses were also of the Cooke Triplet design because my testing showed that by reducing the number of lens elements (pieces of glass) to the minimum I achieved a higher UVA transmission. The lenses were primarily Exakta mount; the prices for fine lenses in this mount are reasonable. Mounting was modified for the long Nikon flange focal distance. In some cases infinity was not possible.

35mm 1:2.8 Tokina Super-Lentar
Exakta and M42 mount
wide angle, excellent UV transmission and little focus shift

35mm 1:2.8 Asunuma Auto-Wide
Nikon F mount
wide angle, uncoated, backup to Lentar

40mm 1:4.5 Schneider-Kreuznach Jsogon
Exakta mount
superb small lens, excellent UV transmission, no focus shift

50mm 1:2.9 Ludwig Meritar
Exakta mount
standard lens for many shots, loved the UV, very little focus shift

80mm 1:5.6 El Nikkor enlarger lens
Leica Thread mount
good on an Exakta double helical extension, good UV transmission but big focus shift

90mm 1:4 Leica Elmar
Leica Thread mount
A softer lens, but good UV transmission, short focus shift

135mm 1:4 Carl Zeiss Jena Triotar
Exakta mount
short telephoto and macro, excellent UV transmission, small focus shift

180mm 1:5.5 Piesker Tele-Picon
superb longer telephoto

Filter
Baader U2 (Venus 2") Ultraviolet Pass Filter
Nikon AF-1 filter holder
Numerous step-rings and spacers

UV Measuring Device
Spectroline DM-365N Long Wave (365nm) Ultraviolet Meter

Accessories
UVP Long-wave Ultraviolet Lamp B-100A with 365nm filter
Reflectors to reduce shadows
Novaflex Bellows for Exakta
Numerous Exakta to Nikon F adapters
Exakta extension tubes
Nikon Extension tubes
Exakta focusing double helical extension
Kalt Copy Stand with lights
4-way focusing rail for copy stand and tripod
Nikon ML-L3 Remote Control
Tripod - generic

Software
PhotoFiltre – 6.3.1
Gimp – 2.4.7
ImageJ – 1.41o

Photographic Procedures

My first attempts at UVR images were done with a Wood's filter that permitted too much infrared light to pass. As a result, the images were contaminated. After learning this, I stepped up to a better camera, lenses, and, most importantly, the Baader U2 filter. None of my early attempts are incorporated in this book.

For many of the photos - such as shots of fly tying materials - the subject is placed on the copy stand and the lens that will provide the optimum size image is mounted on the camera. Since the lenses are manual and the UV Pass filter is opaque to all visible light, all the photography must be done with manual manipulation of the ISO, focus, exposure time, white balance, and aperture. The camera white balance chosen is sometimes critical to the amount of light assigned to each channel. If AUTO is used on the Nikon D40 for UVR, the red channel is saturated and the blue and green are nearly empty. Incandescent 3+ is what I use for internal white balance on the UVR shots.

After focusing, shots are taken in VIS, until a combination of fstop, ISO, and exposure time produces a satisfactory result using available light. Natural light is preferred but not always available.

For UVR I often set the ISO to 800 in order to keep my exposure time down and to allow the lens to be stopped down for a good depth of field. The index chemicals used as UVR indicators must be in the field of each shot; these will be used later for subsequent white balance.

The UV lamp is then started. Since this is a hi-intensity arc lamp it takes 3+ minutes to come to full illumination. The AF-1 filter holder is swung over the lens, putting the Baader filter in place. The Baader filter permits only UV light to pass through and screens out all visible and IR wavelengths (there may be a slight leak at 1500nm IR, but that is still in dispute). Years of optical design went into the construction of this filter which has more than 40 thin coatings preventing all but UV light from passing. Originally designed by astronomers for viewing clouds on Venus, the world of UVR photography was the great winner when this filter became commercially available last year.

Often it is not necessary to adjust the focus used by the VIS shots for the UVR shots - the focus shift is quite minimal with some of the lenses I use. However, for close photos with some lenses, I find it necessary to increase the distance from the subject by two millimeters or more.

The exposure time must be increased for UVR photos. While the VIS shot may use a 1/500th second exposure time, the UVR exposure might be 0.5 seconds to 2.0 seconds. The remote shutter control is convenient, it reduces camera shake dramatically. Camera shake is not as noticeable on the copy stand, but when using a tripod it can be extreme. Patience is required - it might take ten shots to get exposure and UV illumination perfect.

Having captured the images, they are downloaded to a computer for cropping and any other necessary work.

Image Processing

After choosing the best of the VIS and UVR images, they are cropped to remove extraneous background. In the case of the UVR images, the image is first white-balanced, either by hand-leveling the histogram in each channel, or by letting the GIMP2 software perform the leveling automatically. It is important at this point that the index chemicals mentioned previously in the text still appear in the frame as these provide a white as well as a blue/black index for the software to work from.

After white-balancing, the UVR image is often suitable. However, if the VIS and UVR images are complex, it is sometimes necessary to use the UVR image as a filter mask over the VIS image in order to clearly accentuate the differences. In other instances I convert the UVR image to grayscale to highlight the difference.

It is important to understand that the actual information captured by the camera is not altered in the processing of the images. Though colors are changed in the UVR processing, the colors are false to begin with – we have no color for Ultraviolet. What the colors represented, however, which is the degree of UV reflection, does not change. Stretching the histogram allows us to visually distinguish the differences between color values, but does not produce differences that did not exist, nor remove differences that did.

APPENDIX B

UV Reflectance Tables

TABLE X-7

REFLECTION OF WHITE PIGMENTS AND OTHER MATERIALS*

	2537 Å in Percent	2967 Å in Percent	3650 Å in Percent	Visible Light in Percent
Pressed zinc oxide	2.5	2.5	4	88
Barytes	65	70	77	86
Titanium oxide	6	6	31	94
Pressed magnesium oxide	77	86	87	93-95
Smoked magnesium oxide	93	93	94	95-97
Pressed calcium carbonate	78	83	86	96
White wall plaster	46	65	76	90
S.W. white Decotint paint	33	41	58	79
Kalsomine white water paint	12	20	40	70
Albastine white water paint	10	14	45	78
White porcelain enamel	4.7	5.4	63	80
Flat black Egyptian lacquer	5	5	5	5
Five samples of wallpaper	18-31	21-40	33-50	55-75

*M. Luckiesh: Applications of Germicidal, Erythemal and Infrared Energy. New York, D. Van Nostrand Co., 1946, p. 383.

TIO2 Reflectance

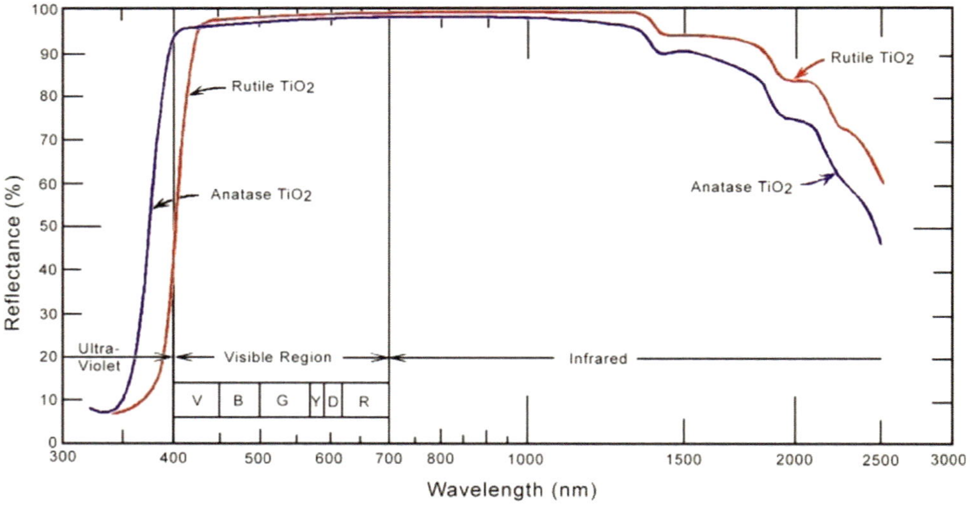

Reflectance of anatase and rutile titanium dioxide pigment through the near UV, visible, and IR regions.

ULTRAVIOLET REFLECTANCE OF DRY WHITE PIGMENTS*

Pigment	Ultraviolet Reflectance Factor in Per Cent
Lead-free zinc oxide	3
35% leaded zinc oxide	4
Zinc Sulfide	6
Titanox B	6
Lead titanate	6
Titanium dioxide	7
Titanox C	7
Lithopone	8
Antimony oxide	17
Zirconium oxide (commercial)	41
Diatomaceous silica (Celite 110)	45
Basic sulfate white lead	48
China clay	54
Aluminum oxide	55
Basic carbonate white lead (Dutch process)	62
Aluminum hydroxide	67
Zirconium oxide, C.P.	78
Magnesium carbonate (commercial)	81

*D. F. Wilcock and W. Soller: Ind. Eng. Chem. 32: 1446, 1940

REFLECTION OF 300nm ULTRAVIOLET FROM THE GROUND

Surface	Per Cent Reflection*
Fresh snow	85
Dry dune sand	17
Sandy turf	2.5
Water	5

K. Buttner and E. Sutter, Strahlentherapie, 54, 156 (1935).

Two reflectors which are common in the laboratory in spectroscopy because of their high reflectance in the visible and the ultraviolet are magnesium oxide and magnesium carbonate. Curves for the reflection of these are shown in Figure X-8 below. The magnesium oxide has a reflectance of over 90 per cent between 260 and 400nm.

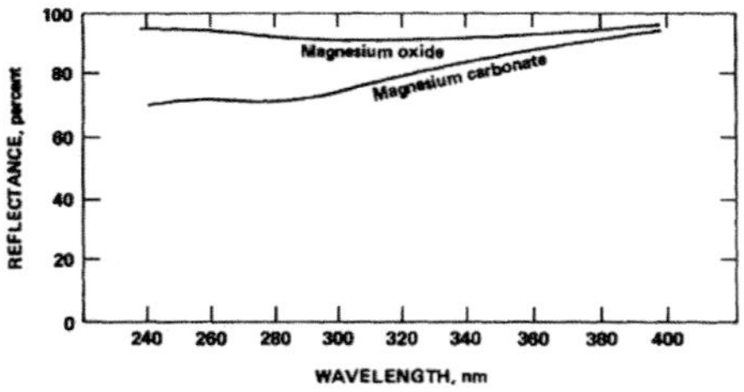

Figure X-8. Reflection from magnesium oxide and magnesium carbonate. From Benford F, Schwartz S, and Lloyd G, J Opt Soc Am 38: 964, 1948.

UV Reflection

Surface	UV Reflectance
Snow	30 – 90%
Ice	7 – 75%
Dry white sand	15 – 30%
White cement	17%
Asphalt	4 – 11%
Water	5 – 10%
Trees	1 – 8%
Grass	1 – 4%

Tevini ed., UV-B Radiation & Ozone Depletion, 1993. p.40

Table 1 Reflectance of ICNIRP/ACGIH-effective solar UV-B from terrain surfaces (Sliney[4])

Representative terrain surfaces	*Diffuse reflectance ACGIH-weighted solar UV-B (%)*
Green mountain grassland	0.8-1.6
Dry, parched grassland	2-3.7
Wooden boat dock	6.4
Black asphalt	5-9
Concrete pavement	8-12
Atlantic beach sand (dry)	15-18
Atlantic beach sand (wet)	7
Sea foam (surf)	25-30
Aged, 'dirty' snow	50
Fresh snow	88

From *Pterygium* By Hugh R. Taylor (Kugler Publications, 2000)

ENDNOTES, CITES, AND ITEMS OF INTEREST

Endnotes

[1] "We previously showed that the brown trout possesses UV-sensitive cones in its retina that are lost in 2-year-old fish. However, present investigations reveal that in the narrow growth zone along the periphery and the ventral embryonic fissure, the formation of these cones continues in trout and salmon."

"Tracts Of Putative Ultraviolet Receptors In The Retina Of The Two-Year-Old Brown Trout (*Salmo Trutta*) And The Atlantic Salmon (*Salmo Salar*)"
Y. W. Kunz
Department of Zoology, University College Dublin, Dublin 4 (Ireland), 2 April 1987

[2] "Ultraviolet visual sensitivity appears to be reduced and, possibly, lost during smoltification in anadromous populations of salmonid fishes. Similar changes occur in non-anadromous salmonids over a mass range that is associated with smoltification in their anadromous conspecifics. However, in sexually mature adult salmonids, ultraviolet-sensitive cones are present in the dorsotemporal retina, suggesting that ultraviolet sensitivity (i) may be regained with sexual maturity or (ii) might never be completely lost… Ultraviolet sensitivity remains intact in the dorsal retina."

"Functional Mapping Of Ultraviolet Photosensitivity During Metamorphic Transitions In A Salmonid Fish, *Oncorhynchus Mykiss*"
Mark E. Deutschlander*, Danielle K. Greaves, Theodore J. Haimberger And Craig W. Hawryshyn
Department of Biology, University of Victoria, PO Box 3020 STN CSC, Victoria, British Columbia, Canada V8W 3N5
Accepted 19 April 2001

[3] "Masu salmon possess well-developed color vision,as shown by the detection of three C-type Spotentials from light-stimulated horizontal cells (Fig. 2). C-type S-potentials have been recorded in several salmonids, including the amago salmon O. masou ishikawae, biwa trout O. masou subspecies, rainbow trout, O. mykiss, brown trout Salmo trutta, and brook trout Salvelinus fontinalis. Microspectrophotometry has been used to demonstrate the presence of three types of cone cell and one type of rod cell in rainbow trout and brown trout, as shown by peak absorbance of different wavelengths of light … Masu salmon are sensitive to ultraviolet light, as shown by electrophysiological recordings of responses to ultraviolet light (Fig. 2) and the histological demonstration that the cone mosaics consisted of three types of cones (Fig. 6a). This result is consistent with that of ultraviolet sensitivity in the behavioral method. The corner single cones of salmonids respond to ultraviolet light. In Japanese dace, ultraviolet sensitivity has been examined using microspectrophotometry and Spotential recording. Recently, ultraviolet sensitivity has been examined in some marine fishes using S-potential recording. The possibility of using ultraviolet light expands the range of wavelengths available to these fishes and may represent an advantage for catching prey or bait, because masu salmon are sight feeders."

Color vision, spectral sensitivity, accommodation, and visual acuity in juvenile masu salmon "Oncorhynchus masou masou"
Norihiko Nakano, Ryo Kawabe, Nariharu Yamashita, Tomonori Hiraishi, Katsutaro Yamamoto and Katsuaki Nashimoto
Fisheries Science
2006; 72: 239–249

[4] "Another interesting tidbit is that juvenile trout possess specialized cones that detect UV light. Early in life a trout's diet consists of zooplankton and other very small creatures. These happen to reflect UV light. These specialized cones help trout locate forage in these early life stages. As the trout begins to grow their diet changes. Studies show that these UV cones are replaced by either rods or cones, helping the trout's vision adjust to the changing diet. These highly specialized cones are usually found in trout younger than 2 years.

A better understanding of optical processing by trout, can only further educate us as anglers upon trout behavior associated with lighting conditions and the aquatic environment.

Not only will we be able to tailor fit our presentations to our lighting scenarios, we will be rewarded with many more fish to hand."

"If trout have the ability to see further into the UV an [sic] IR spectrums than humans can, is it possible that there are colors and materials unnoticeable to us that trout might prefer? It is known that many feathers reflect different colors of light unseen to the human eye. Have you ever wondered why flies that appear to be the same have different affects [sic] on feeding trout?

Perhaps they are tied with similar but not identical materials, emitting many different colors underwater. —Food For Thought—"

from *Seeing the World Through Trout Colored Glasses*
by CFI Fisheries Biologist Toby Stuart
www.cfiglobal.com/documents/2007.pdf

[5]

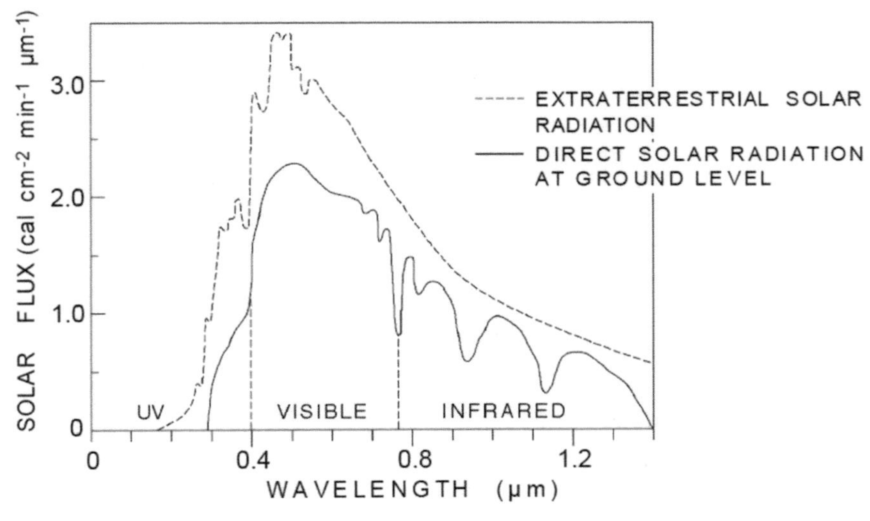

Figure 1. *Spectral composition of solar radiation, extraterrestrially and at ground level. Modified from Wetzel (1983).*

[6] With H. Mueller's proof in 1854 that photoreception occurs in the rods and cones of the eye, steps toward a scientific understanding of human and animal vision began. However, it was not until 1959 that physiological research confirmed that the human retina contains three types of cones, each most responsive to a different wavelength of light, with some overlapping response curves.

[7] "... since the idea that animals may see things that we may not presents a paradox with respect to how scientists go about investigating the sensory world of any animal. These constraints were very much in evidence in the initial efforts concerning the research on ultraviolet-polarization sensitivity in vertebrate animals (Hawryshyn and Beauchamp, 1985). This motivated a series of experiments examining different hypotheses regarding the characteristics of optical stimuli resulting in a clear demonstration of ultraviolet (UV) sensitivity (Hawryshyn and Beauchamp, 1982, 1985). Further experiments confirmed this suggestion that fishes and other vertebrate animals can see UV optical stimuli and that fish have cone photoreceptors that are primarily sensitive to UV light. Initially, ideas such as UV light not penetrating water and UV light being absorbed by the lens of fish eyes provided logical barriers that made UV vision, in any vertebrate, seem improbable. However, these logical barriers turned out to be nothing more than the human eye predicting what the fish could see. More recently, evidence has grown regarding the observation of ultraviolet visual sensitivity in vertebrate animals and the cone photoreceptors that mediate this aspect of vision, dissolving the issue of the scientists's bias (see *Vision Research*, 1994, vol.34, Special Issue on the Biology of Ultraviolet Vision in Animals; *Journal of Experimental Biology*, 2001, vol.204, Special Issue on the Biology of Ultraviolet and PolarizationVision.) from *Mechanisms of Ultraviolet Polarization Vision in Fishes, Craig Hawryshyn, from Sensory Processing in Aquatic Environments* (Springer, NY, 2003)

[8] The normal human eye has a lens containing yellow pigments. These pigments effectively block ultraviolet light from striking the retina. However, those individuals who undergo cataract surgery requiring removal of the lens of the eye are often subsequently able to see in the near ultraviolet. This is a condition known as aphakia. Those with aphakia retain reasonable vision—the cornea of the eye, not the lens, does most of the focusing—and have the newfound ability to see with the ultraviolet light previously blocked by the lens. The retinal sensitivity is such that the blue cones can often detect UVA better than the color blue itself.

One writer postulates that

"An illustration of how ultraviolet appears is provided by the Impressionist painter Claude Monet. Following cataract surgery in 1923, his colour palette changed significantly; after the operation he painted water lilies with more blue than before. This may be because after lens removal he could see ultraviolet light, which would have given a blue cast to the world."
- from www.guardian.co.uk/science/2002/may/30/medicalscience.research

[9]

(19) **United States**
(12) **Patent Application Publication** (10) **Pub. No.: US 2006/0121166 A1**
Jeckle (43) **Pub. Date: Jun. 8, 2006**

(54) **ULTRAVIOLET LIGHT REFLECTIVE COMPOSITION FOR FISHING APPARATUS**

(76) Inventor: **Milan Jeckle**, Spokane, WA (US)

Correspondence Address:
KEITH S BERGMAN
ATTORNEY AT LAW
7 SOUTH HOWARD STREET
SUITE 418
SPOKANE, WA 992013898

(21) Appl. No.: **11/294,086**

(22) Filed: **Dec. 5, 2005**

Related U.S. Application Data

(60) Provisional application No. 60/633,562, filed on Dec. 6, 2004.

Publication Classification

(51) **Int. Cl.**
A23K 1/18 (2006.01)
C09C 1/36 (2006.01)

(52) **U.S. Cl.** **426/395**; 106/436

(57) **ABSTRACT**

A spray-on ultra violet light reflective composition for application on natural and synthetic baits and lures used in fishing activities, and the process for making the composition. The composition is made in three phases each separately mixed and thereafter combined. In phase A, a thickener, optionally a preservative, a pH adjuster and distilled water are combined and mixed. In phase B, a disperser, a solvent, a film forming polymer and a pH buffer are combined and mixed. The mixtures of phase A and phase B are combined, mixed and heated. In phase C, Titanium dioxide, a UV reflectant, a wetting agent and emulsifiers, and optionally a surface smoother, are combined and mixed. The mixture of phase C is heated and added to the heated phase A and phase B mixture. The composition is thereafter cooled, diluted and homogenized.

[10] For example, silver reflects VIS superbly, better than aluminum; however, silver has very poor specular reflective properties in the UV, plummeting from 95% at 370nm to 5% at 300nm, while aluminum continues to reflect well in that range. (J. B. Sabine, Phys. Rev., 55, 1064, 1939.)

[11] "The foundation for the use of reflective ultraviolet photography in circumstances of deep muscle bruising is based on the migration of melanin to the wound's perimeters beneath the skin during the healing process. Melanin absorbs UV light, and even if the wound can no longer be seeen in normal lighting, patterned wounds or injuries will sometimes be detected with UV light long after the initial injury."

Crime Scene Photography By Edward M. Robinson (Academic Press, 2007)

[12] "To analyze the characteristics used by the fish to discriminate the visual training patterns, we performed transfer tests in which the training patterns were replaced by other patterns. The results of all such transfer tests can best be explained by a template-matching mechanism in which the fish stores the view of the rewarded training pattern and chooses from two other patterns the one whose retinal appearance best matches the stored view."

Template-matching describes visual pattern-recognition tasks in the weakly electric fish Gnathonemus petersii
Stefan Schuster* and Silke Amtsfeld
Institut für Biologie I, Hauptstrasse 1, Albert-Ludwigs-Universität Freiburg, D-79104 Freiburg, Germany

[13] "As expected, all 4 species were startled and reacted first to waving rather than food."

The Scotopic Visual Sensitivity Of Four Species Of Trout: A Comparative Study
Russell B. Rader1,2, Timberley Belish1,3, Michael K. Young1,4, and John Rothlisberger5
Western North American Naturalist 67(4), © 2007, pp. 524–537

[14] Although several bird sp ecies have been shown to reflect ultraviolet (UV) light from their plumages, the incidence of UV reflectance, and therefore the potential for UV or UV-enhanced signals, across the avian tree of life is not known. In this study, we collected reflectance data from the plumages of 312 bird species representing 142 families. Our results demonstrate that all avian families possess plumages that reflect significant amounts of UV light. The ubiquity of UV reflectance indicates that all studies of avian behaviour, ecology and evolution involving plumage coloration would benefit from consideration of plumage reflectance in the UV portion of the electromagnetic spectrum. Additionally, we demonstrate the existence of cryptic UV plumage patches and cryptic dimorphism among birds.

The ubiquity of avian ultraviolet plumage reflectance
Muir D. Eaton1,2* and Scott M. Lanyon1,2
1Bell Museum of Natural History, and 2Department of Ecology, Evolution, and Behavior, University of Minnesota,
St Paul, MN 55108, USA

[15] "Because most of the putative UV cones are believed to disappear in early ontogeny, their presence over a large proportion (15-20%) of the surface area of the adult retina suggests that they may be reincorporated prior to or at sexual maturity, at least in rainbow trout." www.ncbi.nlm.nih.gov/pubmed/9184985?dopt=Abstract
"The Contribution of Ultraviolet and Short-Wavelength Sensitive Cone Mechanisms to Color Vision in Rainbow Trout" - David J. Coughlin, Craig W. Hawryshyn - *Brain Behav Evol* 1994;43:219-232

[16] *Degeneration and Regeneration of Ultraviolet Cone Photoreceptors during Development in Rainbow Trout*
Allison, Dann, et al., 2006

[17] *Rainbow trout as a model of retinal photoreceptor death and regeneration*
William Edward (Ted) Allison
B.Sc., University of Victoria, 1996

Also
"The brains of all vertebrates are persistently neurogenic. However, this is not true for the neural

retinas. Only three extant classes of vertebrates show significant posthatch/postnatal retinal neurogenesis: amphibians, birds and fish. The retinas of these animals contain an annulus of progenitors at the margin, from which differentiated neurons emerge. In posthatch amphibians and fish the vast majority of the adult retina is added from the margin and neurogenesis is lifelong, whereas in posthatch birds neurogenesis is limited. Unique to fish, rod photoreceptors are added in situ from stem cells within the mature retina."

Persistent and injury-induced neurogenesis in the vertebrate retina
Peter Hitchcock, a, Malgorzata Ochocinska a, Alexandra Sieh a and Deborah Otteson b
a Department of Ophthalmology and Visual Sciences and The Neuroscience Graduate Program, University of Michigan, Ann Arbor, MI, USA
b Geirrieri Center for Genetic Engineering and Molecular Ophthalmology, Wilmer Eye Institute, Johns Hopkins University School of Medicine, Baltimore, MD, USA 2004

[18] *Male Three-Spined Sticklebacks Reflect In Ultraviolet Light*
Ingolf P. Rick1), Ricarda Modarressie And Theo C.M. Bakker2)
(Institut für Evolutionsbiologie und Ökologie, University of Bonn, An der Immenburg 1, D-53121 Bonn, Germany)

SUMMARY
Recent studies evidenced that some fishes use ultraviolet light in social signalling. Using reflection spectrophotometry we showed that reproductively active male sticklebacks also reflect in the UV-light (300-400 nm). Several body regions showed a reflectance peak in the UV-waveband. Silvery appearing regions reflected stronger than red-coloured or darker ones. Males within one population varied in UV-reflection. The UV-contrast of one particular region was positively correlated with physical condition.
© Koninklijke Brill NV, Leiden, 2004 Behaviour 141, 1531-1541

[19] "In birds, the colors ultraviolet, blue, and green originate from feather microstructure rather than pigmentation. Structural plumage coloration may be maintained by sexual selection, yet there exists only limited evidence that structural colors can honestly signal individual quality. In this study, I found considerable individual variation in the blue-black to ultraviolet coloration of the wing coverts and rump of male Blue-black Grassquits *(Volatinia jacarina),* thereby revealing the potential for sexual selection on structural plumage coloration in this species. I used a principal components analysis (PCA) of four color characteristics to combine reflectance data into a single color score. Birds with high color scores have brighter, more intensely colored, and more saturated blue-black feathers. There was a significant positive relationship between color scores for both body regions and male condition. There was also a positive trend between the color score of the rump region and male body size. These results support recent findings that structural plumage coloration is condition-dependent, and this type of coloration may be an honest signal of male quality in the Blue-black Grassquit."
"Structural Plumage Coloration, Male Body Size, and Condition in The Blue-Black Grassquit"
Stéphanie M. Doucet
Department of Biology, Queen's University, Kingston, ON K7L 3N6, Canada
The Condor 104(1):30-38. 2002

[20] "Several animal species have been shown to use phenotypic traits to assess the competitive ability of opponents and adjust their aggressiveness depending on the likelihood to win the contest. In birds, these phenotypic traits usually involve patches of colored feathers. The benefit to harbor honest signals of male quality is the avoidance of wasteful aggressive interactions. Recent work has shown that ultraviolet (UV) plumage reflectance is an important signal used by females during mate choice. Surprisingly, however, the role of UV signaling on intrasexual selection has been neglected. In the present study, we aimed to test whether UV reflectance of crown feathers was used as a signal of male competitive ability during male-male interactions. Breeding male blue tits *(Parus caeruleus ultramarinus)* were exposed during the female egg-laying period to blue tit taxidermic mounts with either control or reduced UV reflectance of crown feathers. In agreement with the prediction that UV reflectance advertises male quality, we found that breeding blue tits behaved less aggressively toward the UV-reduced decoy. To our knowledge, this is the first experimental evidence suggesting a role for UV signaling on intrasexual selection." –

"Ultraviolet reflectance affects male-male interactions in the blue tit *(Parus caeruleus ultramarinus)*"
Carlos Alonso-Alvarez, 1 Claire Doutrelant, 2 and Gabriele Sorcia, 1
1/Laboratoire de Parasitologie Evolutive, CNRS UMR 7103, Universite´ Pierre et Marie Curie - Paris cedex 05, France, and 2/Laboratoire d'Ecologie Ge´ne´rale CNRS UMR 8571, Muse´um National d'Histoire Naturelle - Brunoy, France
[Behav Ecol 15:805–809 (2004)]

[21] "Female *Bicyclus anynana* butterflies choose males on the basis of their dorsal UV-reflective eyespot pupils" – Kendra A Robertson and Antónia Monteiro – www.pubmedcentral.nih.gov/articlerender.fcgi?artid=1559841

[22] *Communication using eye roll reflective signaling*
I Novales Flamarique, G.A Mueller, C.L Cheng and C.R Figiel
doi: 10.1098/rspb.2006.0246
Proc. R. Soc. B 2007 274, 877-882

[23] "Pure water is transparent to UV wavelengths, but dissolved substances and particulate matter present in natural waters cause significant absorption, reflection, and diffusion of UV within the water column ... Even in the clearest waters, UVB is usually attenuated within the upper 10 m of the water column, although UVB wavelengths have been detected up to 60 m in the very transparent waters of the Southern Ocean during springtime ozone depletion."

[Marine Chemical Ecology, McClintock & Baker, CRC Press, 2001]

[24] "Several species of deep-sea crustaceans possess unusually high spectral sensitivity to UV light, measured electrophysiologically. In addition, behavioral experiments indicate that these species are equally sensitive to near-UV and blue-green light. These results raised questions about the functional significance of this short wavelength sensitivity, since it is commonly believed that UV light is virtually absent in the deep-sea. Using submersible based technology, we conducted measurements of downwelling irradiance at two wavelengths—380 and 480 nm. These measurements indicate that the irradiance of 380 nm light at 500-600 m, the daytime depth range of the crustaceans with a U V sensitivity peak, is high enough to be detected by these species. In addition, measurements of the spectrum of the downwelling light field conducted at dusk demonstrate that spectral changes visible at the surface are not visible at 150 m. These results are discussed with respect to hypotheses on the functional significance of UV vision in deep-sea crustaceans."

UV light in the deep-sea: In situ measurements of downwelling irradiance in relation to the visual threshold sensitivity of UV-sensitive crustaceans.
Authors: Tamara M. Frank a; Edith A. Widder a
Affiliation: a Harbor Branch, Oceanographic Institution, Ft. Pierce, FL
Published in: Marine and Freshwater Behaviour and Physiology, Volume 27, Issue 2 & 3 January 1996, pages 189–197

[25] *Retinal processing and opponent mechanisms mediating ultraviolet polarization sensitivity in rainbow trout (Oncorhynchus mykiss)*

Samuel D. Ramsden1, Leslie Anderson1, Martina Mussi1, Maarten Kamermans2,3 and Craig W. Hawryshyn1,* The Journal of Experimental Biology 211, 1376-1385
Accepted 19 February 2008

The Journal of Experimental Biology 211, 1376-1385
Accepted 19 February 2008

[26] "Data from an alpine stream document for the abundant and widespread mayfly Baetis bicaudatus that, on a continuum from small to large size, the risk of predation increases while the propensity to drift during the day decreases."

Trout Predation and the Size Composition of Stream Drift, by J. David Allan © 1978 American Society of Limnology and Oceanography.

[27] "Diel emergence patterns, swarming and mating behavior, and longevity of seven taxa of adult aquatic insects (Diptera, Ephemeroptera, Trichoptera) from Sycamore Creek, Arizona, were examined. Emergence was synchronized and generally occurred during the 1st 2 hr after dusk; however, the mayfly Tricorythodes dimorphus emerged 3-6 hr later. In the laboratory, adults lived for < 12 hr (Tricorythidae) or > 24 hr (all other taxa) at 25 C. High air temperature (38 C) reduced adult longevity by 47-78%. Mating swarms were observed only at dusk, dawn and early morning. Adult aquatic insects in Sycamore Creek possess life history traits (rapid development, synchronized nocturnal emergence, reduced daytime flight and bimodal adult longevity) that are advantageous in desert streams which periodically dry and flood and are surrounded by a hot, desiccating, riparian habitat containing numerous insectivores."

Diel Emergence, Swarming and Longevity of Selected Adult Aquatic Insects from a Sonoran Desert Stream, by John K. Jackson © 1988
The University of Notre Dame.

[28] *UV matters in shoaling decisions*
Proc. R. Soc. B 2006 273, 849-854
Ricarda Modarressie, Ingolf P Rick and Theo C.M Bakker

[29] "According to our extensive data on Lepidoptera (883 species), UV wing patterns are almost three times more common in nocturnal than in diurnal Lepidoptera. This might be due to predation, because the primary diurnal predators, birds, utilize UV light in foraging and even prefer UV-reflecting prey. To test this hypothesis, we conducted a field experiment with tethered living moths whose wings were artificially manipulated to reflect (UV+, reflection at UV wavelength: 15%) or absorb (UV-) UV light, keeping longer wavelengths identical. Thus, any difference found in survival rates would be the result of the difference in wing patterns in UV spectrum. Significantly more UV+ moths than UV- ones were eaten in the daytime, but no difference in predation rates could be detected when moths were exposed to nocturnal predators. The different survival rates indicate that UV reflection increased predation risk by visually orienting diurnal predators. The lack of difference at night arises from the lack of UV-sensitive predators. UV wing patterns, even if they are important in intraspecies communication, seem to be costly to diurnal Lepidoptera by attracting predators."

"Ultraviolet reflection and predation risk in diurnal and nocturnal Lepidoptera"
Anne Lyytinen, Leena Lindstrom, and Johanna Mappes
Department of Biological and Environmental Science, University of Jyväskylä, P.O. Box 35, FIN-40014, University of Jyväskylä, Finland [Behav Ecol]

[30] *Communication in the ultraviolet*
RE Silberglied - Annual Review of Ecology and Systematics, 1979 - Annual Reviews

[31] Muir et al. (1992 spectral sensitivity) measured several parameters of the compound eye of *Ae. aegypti*. "Spectral sensitivity is the relative sensitivity of the retina to light of different wavelengths and is defined as the inverse of the intensity required to elicit a constant response expressed as a percentage of maximum sensitivity" (Horridge et al. 1975, quoted in Muir (1992)). They found that the retina was sensitive to wavelengths fr om 621 nm to 323 nm (the latter being the lower limit tested). From these measurements it can be concluded that mosquitoes do not see red light. Two sensitivity peaks were found, at 333 and 523 nm in the dorsal part of the eye and at 345 and 523 nm in the ventral part of the eye. These peaks are in the UV (345 nm) and green (523 nm) ranges. Muir et al. suggest that, because other Diptera have photopigments in the UV and green ranges, Ae. aegypti probably also has pigments in these ranges. So far, only one has been isolated, and that at a max of 520 nm (Stein et al. 1979).

From www.ent.iastate.edu/dept/research/vandyk/hostseek.html
From Muir, Lynda E., Thorne, M.J., Kay, B.H. Aedes aegypti (Diptera: Culicidae) Vision: Spectral Sensitivity and Other Perceptual Parameters of the Female Eye - 1992 , J. Med. Entomol.

[32] *Observations on Breeding Behavior of Pachydiplax Longipennis (Odonata :Libellulidae)*
Carol Walser Robey
Museum of Comparative Zoology
Cambridge, Mass. 02138
http://psyche.entclub.org/82/82-089.html

[33] Comparative study of the photopic spectral sensitivity of domestic ducks (Anas platyrhynchos domesticus), turkeys (Meleagris gallopavo gallopavo) and humans
C.L. BARBER1,2, N.B. PRESCOTT1, J.R. JARVIS1, C. LE SUEUR3, G.C. PERRY2 AND C.M. WATHES1
1Silsoe Research Institute, Wrest Park, Silsoe, Bedfordshire, 2Department of Clinical Veterinary Science, University of Bristol, Langford, Bristol, and 3Farm Animal Department, RSPCA, Southwater, West Sussex, England

[34] "Although not all white feathers reflect UV equally (Burkhardt and Finger, 1991), turkey feathers (BUT8) have been shown to be reflective of UV in general (Sherwin and Devereux, 1999) and UV-A in particular (N. S. Hart, School of Biological Science, University of Bristol, Woodland Road, Bristol BS8 1UG, UK, personal communication)."

Effect of Ultraviolet Radiation on the Performance of Intact Male Turkeys
P. D. Lewis,1 G. C. Perry, C. M. Sherwin, and C. Moinard
Department of Clinical Veterinary Science, University of Bristol, Langford, Bristol, BS40 5DU, United Kingdom

[35] CITES (the Convention on International Trade in Endangered Species of Wild Fauna and Flora) is an international agreement between governments. Its aim is to ensure that international trade in specimens of wild animals and plants does not threaten their survival.

[36] This is my theory, a logical extension of the evidence supplied by Varying Hare, Arctic Fox, and Polar Bear fur. There has never been research on this subject.

[37] *Polar Bear Hair Myth*
David B. Schaffer, M.D., Lourene Nevels, PhD., Kimberly Lengel, Senior Curator of Mammals
Philadelphia Zoo, Philadelphia, PA
www.azadocents.org/conferences/2004/polarbear.pdf

[38] "It is a well-documented phenomenon that many aquatic insect species, especially small-bodied Heteroptera (e.g. Sigara, Corixa and Cymatia sp.) but also some Coleoptera (e.g. Berosus sp.), seek new habitats during their migration and dispersal en masse usually between dusk and midnight (e.g. Popham, 1964; Danilevskii, 1965; Johnson, 1969; Fernando and Galbraith, 1973; Zalom et al., 1979, 1980; Danthanarayana, 1986). From an ecological point of view, this is explained conventionally by the reduced risk of both predation and dehydration as well as by the period of calm and lower air temperature at twilight (e.g. Landin, 1968; Landin and Stark, 1973). At sunset, the intensity of ambient light decreases rapidly, rendering the visual detection of flying prey by birds more difficult (e.g. King and Wrubleski, 1998). Since the rate of dehydration is proportional to the surface-to-volume ratio, small-bodied aquatic insects become easily dehydrated during flight if they cannot find a body of water within ~1•h. At nightfall, the lower temperature, higher relative humidity and calmness of air relative to those in daytime are optimal for small-bodied aquatic insects (Landin and Stark, 1973).

...

we conclude that polarotactic water detection is easiest at low solar elevations, because the risk that a polarotactic insect will be unable to recognize the surface of a dark or bright water body is minimal. This partly explains why many aquatic insect species usually fly en masse at dusk. The daily change in the reflectionpolarization pattern of water surfaces is an important visual ecological factor that may contribute to the preference of the twilight period for habitat searching by polarotactic water insects. Air temperature at sunrise is generally low, so dusk is the optimal period for polarotactic aquatic insects to seek new habit ats."

Why is it worth flying at dusk for aquatic insects? Polarotactic water detection is easiest at low solar elevations

Balázs Bernáth1,2, József Gál3 and Gábor Horváth1,*
1Biooptics Laboratory, Department of Biological Physics, Eötvös University, H-1117 Budapest, Pázmány sétány 1, Hungary, 2Plant Protection Institute of the Hungarian Academy of Sciences, Department of Zoology, H-1525
Budapest, P. O. B. 102, Hungary and 3International University Bremen IUB, School of Engineering and Science, P. O. B. 750561, D-28725 Bremen-Grohn, Germany
The Journal of Experimental Biology 207, 755-765
Published by The Company of Biologists 2004

[40] "While colour can be affected by environmental and atmospheric conditions, UV polarization remains stable beneath water or in shady conditions."
http://jeb.biologists.org/cgi/reprint/211/9/ii.pdf

[41] "Regardless of the visual pigments present in the sensory rods and cones, wavelengths that are completely absorbed or reflected by the pre-retinal ocular media cannot be perceived.

It is reasonable to hypothesize that by determining which portions of the available radiation are absorbed prior to reaching the retina, we would gain a strong indicator of the visual pigments that are not likely to be found in the retina(Douglas and McGuigan 1989). Douglas and McGuigan (1989) indicated several groups of freshwater fishes that were unlikely to possess ultraviolet (UV)-sensitive visual pigments, since most of the UV radiation was absorbed by the lens. Detection of a visibly "yellow" lens cannot, however, be used as evidence of a lack of UV vision, since some yellow pigments do not absorb in the critical UV wavelengths (Kennedy and Milkman 1956), or are present in low densities and allow sufficient transmission for vision at other wavelengths. More recent studies of both shallow- and deepsea fishes have revealed a wide range of absorption spectra, with UV-blocking lenses especially common in advanced perciform fishes (Douglas and McGuigan 1989; Douglas and Thorpe 1992; Douglas et al. 1995; Dunlap et al. 1989; Thorpe and Douglas 1993; Thorpe et al. 1993; Siebeck and Marshall 2001)."

Ultraviolet blocking in the ocular humors of the teleost fish Acanthocybium solandri (Scombridae)
Peter A. Nelson, Jill P. Zamzow, and George S. Losey

[42] "The skin of fish is often richly coloured and has compositions and patterns of a striking elegance, like for example those of tropical fish. The white and the iridescence of Fish are due to deposits of small guanine crystals, and the ventral part is particularly rich in these. The phenomenon of iridescence and the shiny effect of the birifringent crystals can be modified by changes of expansion and concentration of electro-active melanin inside the chromatophore. The blue colour of Fish can be due either to a chromoprotein or a schemochrome, the green to the effect of a schemchrome with a yellow pigment, while the dark green and the olive green come from the combination of a black and a yellow or orange chromatophore. The colouration of Fish is often in relation to their physical and psychic state or to their environments. The animals can change colouraration quickly and notably under the action of physical stimuli by a neurohormonal mechanism. In some species of Fish, often in those of the abysses, the skin has luminous organs of a glandular nature which emit light, either spontaneously or under the action of physical stimuli, and which can be switched on or off. The light may be coloured because of the presence of chromatophores on the derma which cover the generator organs of light and takes on yellow, red, blue colours with an intensity which varies, sometimes from individual to individual.

The pigments which contribute to the colouration of Fishes belong to the classes of melanins, carotenoids, pterines, flavines, bile pigments and purine or guanine bases (white)."
from www.tightrope.it/nicolaus/origins.htm

[43] "Pigments are compounds that absorb particular wavelengths of light and can contribute to the color of biological patches. Two classes of pigments that are commonly studied in fish are carotenoids (usually yellows, orange, and red) and melanin (browns, blacks, and grays); they are deposited in the integument. Short wavelength (blue and violet) and silvery coloration in vertebrates are almost always structurally based, the result of selective light scatter owing to variable refraction within the integument, and only one blue pigment has been described in fish (reviewed in Bagnara et al.2)."

Pigments, Patterns, and Fish Behavior
Anna C. Price, Cameron J. Weadick, Janet Shim, and F. Helen Rodd
ZEBRAFISH
Volume 5, Number 4, 2008
Copyright, Mary Ann Liebert, Inc.

[44] "Interestingly, ambient light at dawn and dusk contains a particularly high proportion of short wavelengths."

Bat Eyes Have Ultraviolet-Sensitive Cone Photoreceptors
Müller B, Glösmann M, Peichl L, Knop GC, Hagemann C, et al. 2009
PLoS ONE 4(7): e6390. doi:10.1371/journal.pone.0006390

[45] "The second experiment investigated more specifically whether UV cues are especially important in dawn and dusk when short wavelengths are high in the proportion of available light."

Do house mice use UV cues when foraging?
Johanna Honkavaara, Helena Åberg and Jussi Viitala
Journal of Ethology
Volume 26, Number 3 / September, 2008

Cites and items of interest

"We previously showed that the brown trout possesses UV-sensitive cones in its retina that are lost in 2-year-old fish. However, present investigations reveal that in the narrow growth zone along the periphery and the ventral embryonic fissure, the formation of these cones continues in trout and salmon."

"Tracts Of Putative Ultraviolet Receptors In The Retina Of The Two-Year-Old Brown Trout *(Salmo Trutta)* And The Atlantic Salmon *(Salmo Salar)*"
Y. W. Kunz
Department of Zoology, University College Dublin, Dublin 4 (Ireland), 2 April 1987

"Ultraviolet visual sensitivity appears to be reduced and, possibly, lost during smoltification in anadromous populations of salmonid fishes. Similar changes occur in non-anadromous salmonids over a mass range that is associated with smoltification in their anadromous conspecifics. However, in sexually mature adult salmonids, ultraviolet-sensitive cones are present in the dorsotemporal retina, suggesting that ultraviolet sensitivity (i) may be regained with sexual maturity or (ii) might never be completely lost… Ultraviolet sensitivity remains intact in the dorsal retina."

"Functional Mapping Of Ultraviolet Photosensitivity During Metamorphic Transitions In A Salmonid Fish, *Oncorhynchus Mykiss*"
Mark E. Deutschlander*, Danielle K. Greaves, Theodore J. Haimberger And Craig W. Hawryshyn
Department of Biology, University of Victoria, PO Box 3020 STN CSC, Victoria, British Columbia, Canada V8W 3N5 Accepted 19 April 2001

"In order to facilitate emerging models of retinal development, we developed electroretinogram and in situ hybridization protocols to examine the ontogeny of photoreceptors in the retina of a land-locked salmonid, the rainbow trout (Oncorhynchus mykiss). We cloned cDNA fragments corresponding to the rod opsin and each of the four cone opsin gene families, which we utilized to produce riboprobes. We established the specificity of the in situ hybridization protocol by examining subcellular signal localization and through double-labeling experiments. We confirm the assumption that the accessory corner cones in the square mosaic are the ultraviolet wavelength-sensitive (UVS) cone photoreceptor (i.e., they express an SWS1 opsin) and observed UVS cones throughout the retina of small trout. Larger fish have a decrease in sensitivity to short wavelength light stimuli and the distribution of UVS cones in the mature retina is limited to the dorsal-temporal quadrant. These larger fish also possess differentiated UVS cones in the peripheral germinal zone (PGZ), including within areas peripheral to mature retina lacking UVS cones. These data are consistent with the loss of putative UVS cones from the PGZ of a migratory salmonid of another genus, and thus the disappearance of UVS cones appears to be general to the Family Salmonidae, regardless of life history strategy. The generation, differentiation, and subsequent loss of UVS cones in the smolt PGZ is a dramatic example of the supposition that the mechanisms of PGZ development recapitulate the retinal embryogenesis of that species."

Ontogeny of ultraviolet-sensitive cones in the retina of rainbow trout (Oncorhynchus mykiss)
W. Ted Allison 1, Stephen G. Dann 1, Jon Vidar Helvik 2, Clarrisa Bradley 1, Heather D. Moyer 1, Craig W. Hawryshyn 1 *
1Department of Biology, University of Victoria, Victoria, British Columbia V8W 3N5, Canada
2Department of Molecular Biology, University of Bergen, 5020 Bergen, Norway
email: Craig W. Hawryshyn (chawrysh@uvic.ca)
*Correspondence to Craig W. Hawryshyn, Department of Biology, University of Victoria, Victoria, British Columbia V8W 3N5, Canada
J. Comp. Neurol. 461:294-306, 2003. © 2003 Wiley-Liss, Inc.

"At dusk, ambient light is UV blue with dominant wavelengths between 345 and 480 nm peaking at 405 nm (Figure 2c). Forest shade is greenish with a peak at 550 nm. Woodland shade is bluish with an irradiance spectrum peaking at 455 nm. Small gaps are yellow-orange, dominated by long wavelengths. Large gaps are whitish with a flatter spectrum in the visible wavelengths. Mean light intensity is very low at dusk (1.0 lmol/m2 s between 300 and 700 nm), compared with forest shade (8.6 lmol/m2 s), woodland shade (25.0 lmol/m2 s), small gaps (188.8 lmol/m2 s), and large gaps (1302.9 lmol/m2 s)."
"The study was conducted in undisturbed primary tropical rain forest at the Nouragues Centre National de la Recherche Scientifique Biological Station in French Guyana (4_03#N, 52_42#W; Bongers et al. 2001)."
Dusk light environment optimizes visual perception of conspecifics in a crepuscular horned beetle
Marc The´ry,a Sylvain Pincebourde,b and Francxois Feera
Behavioral Ecology

"Evidence that ultraviolet markings are associated with patterns of molecular gene flow" - Roger S. Thorpe and Murielle Richard
www.pnas.org/cgi/content/full/pnas;98/7/3929

"Many fishes, amphibians, reptiles, birds, and some mammals use UV vision for such basic activities as foraging (Burkhardt, 1982; Viitala et al., 1995) and mate choice (Bennett et al., 1996, 1997). These species detect light maximally at 360–370 nm by using UV pigments."
From *Molecular evolution of color vision in vertebrates*
Shozo Yokoyama
Department of Biology, Biological Research Laboratories, Syracuse University, 130 College Place, Syracuse, NY 13244, USA
17 July 2002

"The data from the trout confirm that the tectum plays an important role in visual functions such as acuity, pattern vision and vision of movement. The common feature of habituation in the torus suggests that the torus may be involved in the signalling of sudden changes in the visual scenery." From *Response Features Of Visual Units In The Lower Midbrain Of The Rainbow Trout*
By Nico A. M. Schellart And Walther E. I. Rikkert
The Company of Biologists Limited 1989

Tom Rosenbauer in his book *The Orvis Guide to Prospecting for Trout* (Globe Pequot, NY, 2000), mentions the Parachute Hare's Ear, as his favorite searching dry fly. Essentially, this fly is a Parachute Adams with all grizzly hackle – no brown – and a fuzzier body. We know how the Parachute Adams lights up with UVR, could this be what makes the Parachute Hare's Ear so effective?

"These measurements were performed on museum skins held by the Queensland Museum and on live specimens from Currumbin Sanctuary on the Gold Coast, Australia. Most of these reflectance spectra have been published previously, along with a detailed methodology (Vorobyev et al. 1998). We also conducted a study to examine the effects of specimen age on fading in museum samples and found that, although some fading does occur, the main change is a loss of 'chroma' and 'brightness' with little change in 'hue' (Hausmann 1997; McNaught

& Owens 2002; J. Marshall, unpublished data; and see Endler (1990) for terminology). In addition, the UVreflective component of colours does not change differentially relative to the rest of the colour spectrum during fading. We were able to develop, therefore, methods of scoring the presence or absence of UV reflectance and fluorescence that were not affected by fading."

Proc. R. Soc. Lond. B 2003 270, 61-67

Franziska Hausmann, Kathryn E. Arnold, N. Justin Marshall and Ian P. F. Owens

The electromagnetic spectrum of ultraviolet light can be subdivided in a number of ways. The draft ISO standard on determining solar irradiances (ISO-DIS-21348)[2] describes the following ranges:

Name	**Abbreviation**	**Wavelength range in nanometers**	**Energy per photon**
Ultraviolet A, long wave, or black light	UVA	400 nm - 315 nm	3.10 - 3.94 eV
Near	NUV	400 nm - 300 nm	3.10 - 4.13 eV
Ultraviolet B or medium wave	UVB	315 nm - 280 nm	3.94 - 4.43 eV
Middle	MUV	300 nm - 200 nm	4.13 - 6.20 eV
Ultraviolet C, short wave, or germicidal	UVC	280 nm - 100 nm	4.43 - 12.4 eV
Far	FUV	200 nm - 122 nm	6.20 - 10.2 eV
Vacuum	VUV	200 nm - 10 nm	6.20 - 124 eV
Extreme	EUV	121 nm - 10 nm	10.2 - 124 eV

ADDITIONAL THANKS TO...

Dan Curry for his unwavering support.

Eric Reaves, for his fine drawings and cartoons. Eric has been drawing "Garfield" comic strips for the creator Jim Davis for fifteen years. He feels the need to balance the cat with some fish occasionally. More of his work may be seen at www.ericreavesart.com.

Marcos Vergara, of Hareline Dubbin (www.harelinedubbin.com), who saw the angling possibilities of reflected UV and took action. Through his assistance I had the materials I needed for my work.

Eric Austin, who tied most of the flies used in this book. Eric's flies are almost too beautiful to fish. More examples of Eric's work may be found at www.traditionalflies.com.

Dr. Klaus D. Schmitt, for the kind advice on appropriate tools for UVR photography.

Bjørn Rørslett, for his generous advice, his highly informative website (www.naturfotograf.com), and his groundbreaking work in UV photography.

CPSIA information can be obtained
at www.ICGtesting.com
Printed in the USA
LVIC060702240413
330671LV00006B
* 9 7 8 0 9 8 4 0 8 6 3 0 6 *